Pemberton

Mathematics

for Cambridge IGCSE®
Teacher Resource Pack

Third edition

For the updated syllabus

Extended

Deborah Barton

Oxford excellence for Cambridge IGCSE®

OXFORD
UNIVERSITY PRESS

OXFORD
UNIVERSITY PRESS

Great Clarendon Street, Oxford, OX2 6DP, United Kingdom

Oxford University Press is a department of the University of Oxford. It furthers the University's objective of excellence in research, scholarship, and education by publishing worldwide. Oxford is a registered trademark of Oxford University Press in the UK and in certain other countries

First published 2012

Second edition 2017

Third edition 2019

British Library Cataloguing in Publication Data

Data available

ISBN: 978-0-19-842847-3
10 9 8 7 6 5 4 3

Printed by CPI Group (UK) Ltd, Croydon CR0 4YY

Acknowledgements

Cover photo by Shutterstock

Paper used in the production of this book is a natural, recyclable product made from wood grown in sustainable forests. The manufacturing process conforms to the environmental regulations of the country of origin.

About this book

This book is designed to provide teachers with useful resources to support teaching the extended syllabus for IGCSE® Mathematics. The activities are ordered to match the contents of the Oxford University Press book *Pemberton Mathematics for Cambridge IGCSE® Extended*, though they can successfully be used with other courses. I hope that your students will find these resources both enjoyable and extremely useful.

The resources included in this book promote active learning; this ensures better understanding, and consequently better retention of the subject knowledge. There is plenty of scope for discussion work and group activities, both of which are valuable tools for reinforcing learning. Often students complete such activities thinking that they have done very little work as they have written little or nothing down. However, the work completed is often of greater value than a written activity, as there is more scope for sharing knowledge and eliminating common misconceptions than when students work individually. There is less pressure than in a written exercise as many of these resources can have different starting points, meaning students can work in an order that suits them and their level of understanding.

Teachers are sometimes concerned about using these types of resource because of various factors, including the increased volume of classroom noise, the increase in preparation time and the perceived lack of teacher control. It is worth overcoming these issues, as active learning is definitely beneficial for students. In many of these activities there are no limits to learning, particularly when using the creative alternative approaches. You will find many students work at higher levels than you expect, because of their natural competitiveness and desire to improve.

Also, in group work and discussion students are required to articulate sometimes quite complex mathematical ideas that wouldn't be covered when working individually. Not everyone learns best by writing things down. These activities address alternative learning styles particularly if you find your classes are often the "teacher gives examples/students do an exercise" approach.

Many of the resources include jigsaw pieces or card sorting activities. You could cut these out yourself beforehand to save lesson time. Alternatively you could ask students to cut out resources themselves which not only saves your time, but also gives them the opportunity for some initial thinking time or discussion time.

Each resource is split into the following headings.

Teacher notes

This section gives an outline of the activity, what students need to know to complete the activity and some suggestions of how the activity can be used.

Introductory activity

There is a suggested introductory activity for each resource to help you and your students. This section often includes some helpful examples to go through with your students before beginning the activity. Alternatively, there may simply be a suggestion to use this after a particular chapter in the book, leaving the introduction to a minimum in order to ensure the activity is more challenging for the students.

Description of activity

This section includes a description of how to use the resource and what other resources or photocopying you will need, along with other helpful suggestions and hints.

Differentiation

All good resources should provide you with the opportunity to simplify or extend the resource as necessary. This section explains different ways that the resource can be adapted to cater for your less able or more able students.

Alternative approaches

Sometimes an activity takes less time than you anticipate and this can be difficult to manage in lessons. This section suggests some alternative ways to use the resource, and to extend it, which is always useful! Your "early finishers" will be extended appropriately rather than given more of the same (which often frustrates them).

Answers

All resources include answers. I would suggest keeping a photocopy of these answers with each prepared resource.

Photocopiable resources

For activities such as jigsaws and card sorts, several sets of resources will be needed, one for each group. In this case I recommend using different coloured card for each set so that any dropped or lost pieces can be easily identified and returned to the correct pack. For most of these activities groups are generally better with no more than four students, as discussions can be less inclusive in larger groups, but obviously this depends on your students. Zip-lock bags or envelopes are also very useful for storing the card pieces, and laminating them will make them last longer. In some cases you may choose to print the resource on paper rather than card, so that students can stick the completed jigsaw into their books. You do not have to use the resources for the entire class and may choose to use some resources for one or two individuals instead.

Online resources

The online components are designed to supplement and support the teaching of this course. There are blank activity sheets, presentations, and a scheme of work.

Access your support website at www.oxfordsecondary.com/9780198428473

Finally, I hope that you enjoy using these resources and that your students find them beneficial as well as fun. I also hope these resources help your students really enjoy maths!

Deborah Barton

Contents

Multiples, primes, squares, factors, cubes and roots crossnumber

Teacher notes

This activity is for students who already know about multiples, factors, primes, squares, cubes and roots.

It is meant to be a fun revision session or a quick homework activity. In this activity students will need to know where the relevant square, cube, square root and cube root keys are on their calculators. They will also need to be able to understand how to solve simple equations such as $x^2 = 361$ to find x. This activity also covers some simple substitution and students will need to understand the word product. Students also need to be able to work out highest common factors and lowest common multiples.

Introductory activity

There should be little need for any introduction, however you may want to remind students how to tackle solving equations such as:

If $\sqrt[3]{x} = 13$ find x students need to do $13^3 = 2197$

If $x^2 = 289$ find x students need to do $\sqrt{289} = 17$

You may (or may not) also wish to remind students that 1 is not a prime number as this is a common misconception. Students occasionally get mixed up between factors and multiples so you could also remind them about these.

Crossnumber activity

Print out one copy of the crossnumber grid and clues for your students, or pairs of students, ask them to complete the crossnumber by filling in one digit per box.

Differentiation

Ask students to race to complete the crossnumber. Alternatively use the blank grid, from the online resources, to make harder/easier clues of your own.

Alternative approaches

Give the students a blank copy of the crossword and ask them to write both the clues and the solutions; this is quite challenging.

Answers

1 1	4	**2** 4		**3** 5	1	**4** 2		**5** 1	2
		4		8		**6** 4	0	5	
	7 2	5		**8** 3	1			5	
	0			2				6	
	9 1	**10** 2				**11** 1	3	3	**12** 1
		8		**13** 2	**14** 1	0			2
		0			0		**15** 2	4	9
16 7	2	9			2				6
0				**17** 9	4	0	9		

Multiples, primes, squares, factors, cubes and roots crossnumber

1		2		3		4		5	
						6			
	7			8					
	9	10				11			12
				13	14				
							15		
16									
				17					

Across

1 12^2

3 8^3

5 The highest common factor of 24 and 36

6 $\sqrt{164025}$

7 A multiple of 5, also a square number

8 A prime number between 30 and 36

9 $\sqrt[3]{1728}$

11 $(a + b)^3$ where $a = -2$ and $b = 13$

13 The product of the first four prime numbers

15 $3 + 3^2 \times 3^3 + 3$

16 $(3^3)^2$

17 (The largest prime number under $100)^2$

Down

2 $21^2 + 2 \times$ (the first prime number)

3 $\sqrt[3]{x} = 18$, find x

4 The largest multiple of 8 under 30

5 $25^3 - 8^2 + 2^x$ where x is a factor of every number

7 $x^2 = 40401$, find x

10 (The smallest prime number over $50)^2$

11 The largest factor of 10

12 $(6^2)^2$

14 $4^3 \times 4^2$

16 The lowest common multiple of 7 and 10

Decimal places and significant figures hidden treasure hunt

Teacher notes

This activity can be used as:

- A starter: an introduction to the topic for students who have studied this topic in a lower year group.
- A plenary: an end of lesson revision/recap session after having completed the work on significant figures and decimal places.
- A revision exercise later on in the year.

Often students will miss out on marks during an exam due to rounding errors. On the front of the exam paper it states "If the degree of accuracy is not specified in the question, and if the answer is not exact, give the answer to three significant figures. Give answers in degrees to one decimal place" There are a large number of students who ignore these instructions (generally they over-round) and lose valuable marks. Since rounding is studied very early in the course it might be a good idea to use this activity close to the exams to reinforce the importance of rounding.

Introductory activity

Begin by writing the following statement on the board: "If the degree of accuracy is not specified in the question, and if the answer is not exact, give the answer to three significant figures. Give answers in degrees to one decimal place." Also point out to students that only the **final answer** should be rounded and rounding should not be done part-way through a calculation. Many marks are lost due to "premature rounding".

Hidden treasure hunt game

On the board draw a set of co-ordinate axes with both the x- and y-axes from 0 to 3.

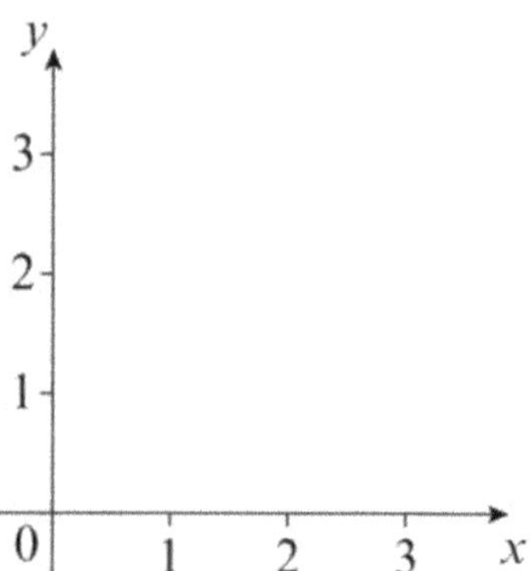

Have a list of questions on rounding prepared. For example you could use:

(0, 0)	4.5 to nearest whole number	(2, 0)	124.678 to 3 s.f.
(0, 1)	6.2431 to 2 d.p.	(2, 1)	4.3971 to 2 d.p.
(0, 2)	23.449 to 1 d.p	(2, 2)	758 to nearest ten
(0, 3)	74.358 to 3 s.f.	(2, 3)	0.0124 to 2 d.p.
(1, 0)	0.0023456 to 3 s.f.	(3, 0)	5.9999 to 1 d.p.
(1, 1)	0.059 to 1 d.p	(3, 1)	754821 to 3 s.f.
(1, 2)	0.084321 to 3 s.f.	(3, 2)	25.321 to 1 d.p
(1, 3)	0.789 to nearest whole number	(3, 3)	4569 to nearest hundred

Explain that this is a treasure map and somewhere at one of the co-ordinates there is buried treasure. Their aim is to try to find the treasure.

Invite students in turn to state co-ordinates and then answer the associated question.

If they get the answer right, tell the class if the treasure is buried there or not and mark off on the axes with a cross where their guess was.

If they get the answer wrong, do not tell them if the treasure is buried there or not and the class loses a life. The class has five lives.

Differentiation

You could reduce the number of lives to make this harder or increase the number of lives to make it easier. You may choose not to mark off where the students have guessed. Then the students will have to pay close attention to where other students have already guessed or they will end up answering the same questions again.

Alternative approaches

Choose your own questions and write them in a copy of the blank grid, available from the online resources.

Answers

(0, 0)	4.5 to nearest whole number	5	(2, 0)	124.678 to 3 s.f.	125
(0, 1)	6.2431 to 2 d.p.	6.24	(2, 1)	4.3971 to 2 d.p.	4.40
(0, 2)	23.449 to 1 d.p	23.4	(2, 2)	758 to nearest ten	760
(0, 3)	74.358 to 3 s.f.	74.4	(2, 3)	0.0124 to 2 d.p.	0.01
(1, 0)	0.0023456 to 3 s.f.	0.00235	(3, 0)	5.9999 to 1 d.p.	6.0
(1, 1)	0.059 to 1 d.p	0.1	(3, 1)	754821 to 3 s.f.	755 000
(1, 2)	0.084321 to 3 s.f.	0.0843	(3, 2)	25.321 to 1 d.p	25.3
(1, 3)	0.789 to nearest whole number	1	(3, 3)	4569 to nearest hundred	4600

Solving linear equations jigsaw

Teacher notes

This jigsaw activity is an interesting group work alternative to a written exercise. The advantages of a jigsaw activity over a written exercise are that it creates opportunities for students to discuss their ideas, and they can find an alternative starting point if they are not sure how to do a particular question. Students are able to start with the answer and work backwards, which is sometimes an interesting way to tackle a question.

This activity requires students to be able to solve both simple equations and more complex linear equations involving brackets and unknowns on both sides of the equals sign.

Introductory activity

Students generally have the most success solving linear equations using the balance method and good clear layout of solutions is a key part of maximising method marks. Some students like to use a flow chart method for solving equations, however they frequently lose method marks and struggle when there are unknowns on both sides of the equation.

If you want to use this task as a revision activity, or a plenary, you may find that little introduction is necessary. Otherwise you may want to go through the following examples.

Example 1

Solve $3x + 20 = 2$ — the first step is to subtract 20 from both sides

$3x = -18$ — then divide both sides the coefficient of x which is 3

$x = -6$

Example 2

Solve

$3(2x + 5) = 5(2x - 1)$ — multiply out brackets

$6x + 15 = 10x - 5$ — – $6x$ from both sides

$15 = 4x - 5$ — + 5 to both sides

$20 = 4x$ — then divide both sides by the coefficient of x which is 4

$5 = x$ — this is the same as $x = 5$, and it is acceptable to leave it in this form

Usually the answer line has $x =$ ………, then students can just write 5 in the space. If it doesn't have the $x =$ … part printed they must write $x = 5$.

Example 3

Solve $3 + \frac{x}{5} = 1$ — subtract 3 from both sides

$\frac{x}{5} = -2$ — multiply both sides by the divisor 5 to clear the fraction

$x = -10$

Example 4

Solve

$20 - 3(4x - 2) = 5(2x + 3)$ — multiply out brackets taking care with the minus signs

$20 - 12x + 6 = 10x + 15$ — simplify

$26 - 12x = 10x + 15$ — + $12x$ to both sides

$26 = 22x + 15$ — – 15 from both sides

$11 = 22x$ — divide both sides by the coefficient of x which is 22

$0.5 = x$

A common error here is to have the answer $x = 2$ because students like to have a whole number answer. Remind them to divide by the coefficient of x whether it is on the LHS or the RHS.

Jigsaw activity

Ask your students to work in small groups. Copy the two pages of jigsaw pieces onto card to make a complete set for each group. There are several different-shaped jigsaws used in the Teacher Resource Kit so you should not tell the students what shape to expect. This jigsaw has no blank edges so it is harder for your students to complete.

Differentiation

To make it more difficult there are questions or 'solutions' around the outside edges of the jigsaw. If you want to make it easier, you could indicate outside edges; perhaps by drawing a bold line along the outside edges.

Alternative approaches

You could just ask students to complete the jigsaw or you could make use of the fact that the outside edges are not blank. You could copy blank jigsaw pieces from the online resources and ask students to use these pieces to extend the jigsaw: either finding the solutions to the outside questions or writing a question that has the given solution. Alternatively students could make an entirely new jigsaw of their own.

Answers

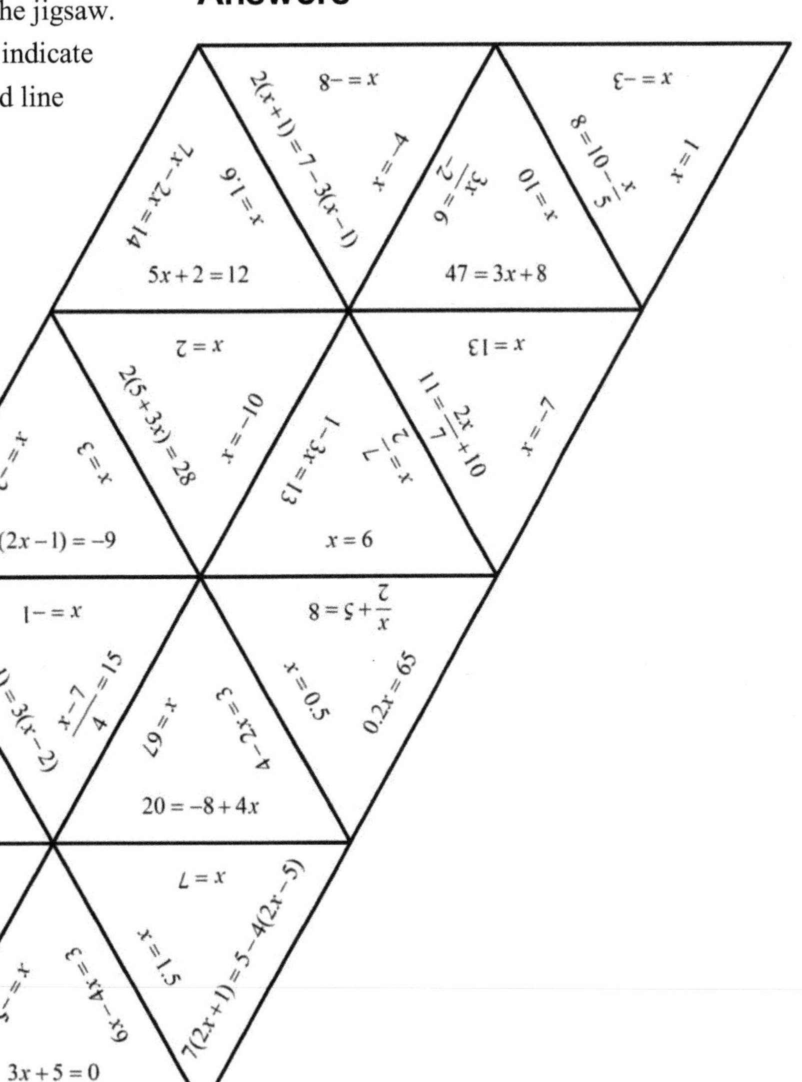

Solving linear equations jigsaw pieces

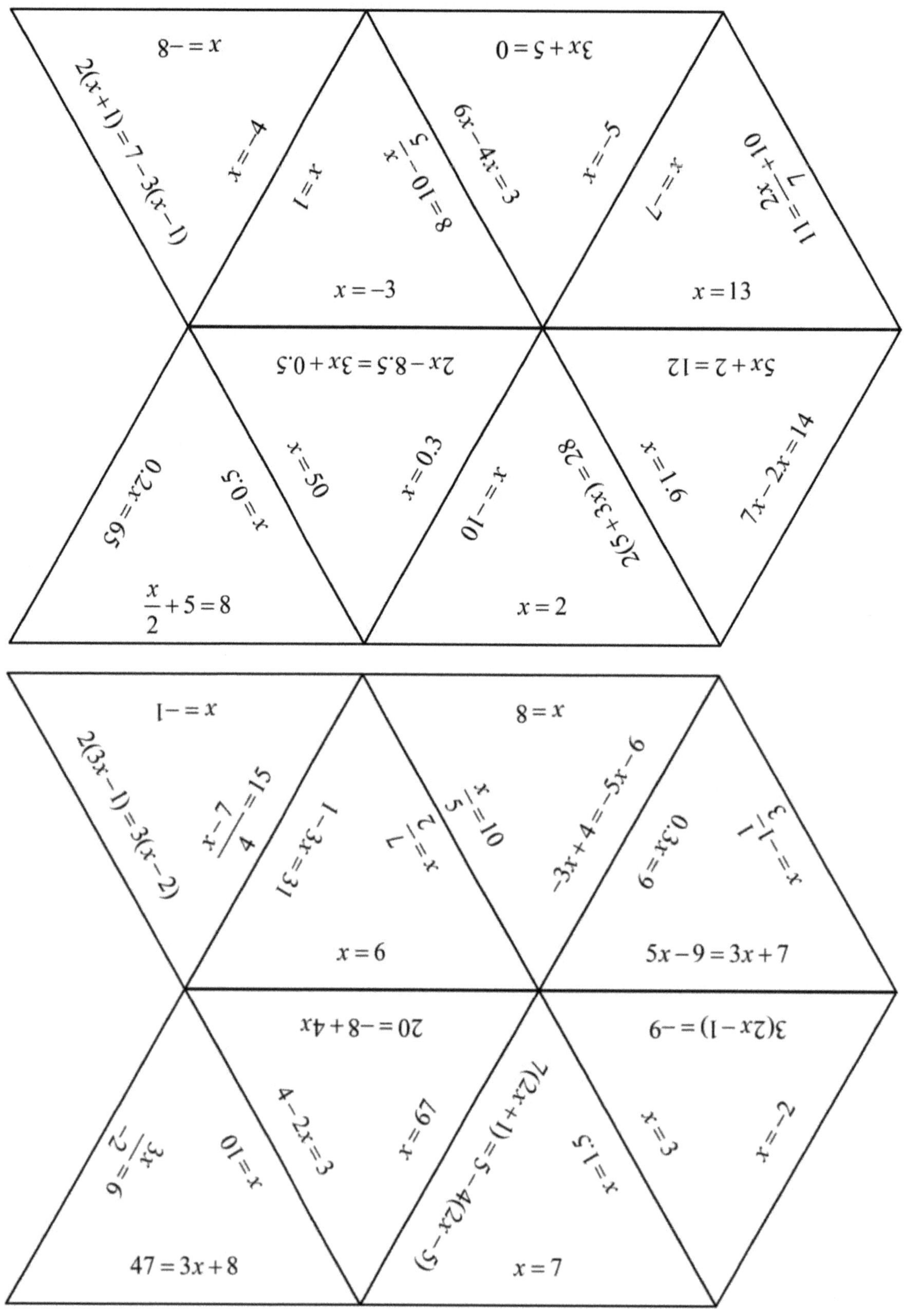

Larger size jigsaw pieces are available in the online resources.

Polygons, angles and symmetry top trumps

Teacher notes

This top trumps game is an interesting and fun revision activity covering regular and irregular polygons; interior and exterior angles; reflectional and rotational symmetry; vocabulary of 2D and 3D shapes; angles in parallel lines and planes of symmetry.

Introductory activity

If you want to use this task as a revision activity or a plenary you may find that little introduction is necessary. Otherwise you may want to go through some of the key vocabulary or topics:

- Regular polygon
- Interior angle
- Exterior angle
- Order of rotational symmetry
- Lines of reflectional symmetry
- Corresponding angles
- Alternate angles
- Vertically opposite angles
- Planes of symmetry
- Vertices, edges, faces

Top trumps activity

Ask your students to work in pairs. All of the cards in the pack are dealt out evenly between the pairs of students. The starting player chooses a category from their top card and reads out its value. The numbers corresponding to each category will need to be worked out by the player. The other player then works out and reads out the value of the same category from their top card. The largest value wins and the winner takes both the cards and places them at the bottom of their pile. The winning player then chooses the category for the next round. If both cards have the same value they are placed in the centre and a new category is chosen from the next card by the same person as in the previous round. The winner of that round obtains all of the cards in the centre as well as the top card from each player. The winner of the game is the player who obtains the whole pack of cards.

Differentiation

To make it easier you could allow students to work out all the angles before playing the game (as the order you work out some of the angles could make other angles easier). You could allow students to write on their set of cards.

To extend the activity you could ask students to come up with a fifth category for each card and an associated angle fact to go with it.

Alternative approaches

There are some blank cards included in the online resources that students could use to create their own top trump cards to add to these ones.

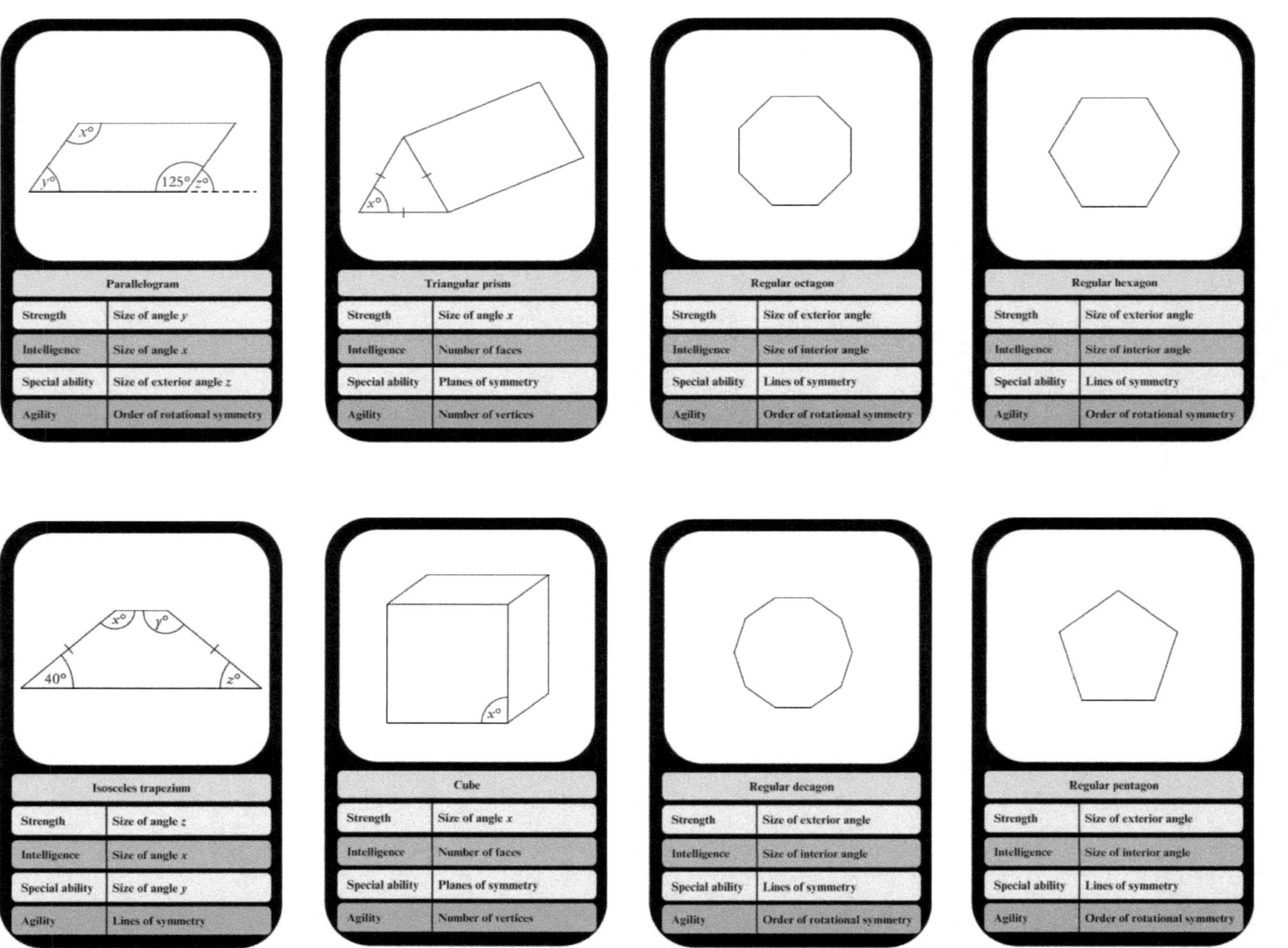

x°
y°
125°
z°
Parallelogram
Strength
Size of angle y
Intelligence
Size of angle x
Special ability
Size of exterior angle z
Agility
Order of rotational symmetry
x°
Triangular prism
Strength
Size of angle x
Intelligence
Number of faces
Special ability
Planes of symmetry
Agility
Number of vertices
Regular octagon
Strength
Size of exterior angle
Intelligence
Size of interior angle
Special ability
Lines of symmetry
Agility
Order of rotational symmetry
Regular hexagon
Strength
Size of exterior angle
Intelligence
Size of interior angle
Special ability
Lines of symmetry
Agility
Order of rotational symmetry
x°
y°
40°
z°
Isosceles trapezium
Strength
Size of angle z
Intelligence
Size of angle x
Special ability
Size of angle y
Agility
Lines of symmetry
x°
Cube
Strength
Size of angle x
Intelligence
Number of faces
Special ability
Planes of symmetry
Agility
Number of vertices
Regular decagon
Strength
Size of exterior angle
Intelligence
Size of interior angle
Special ability
Lines of symmetry
Agility
Order of rotational symmetry
Regular pentagon
Strength
Size of exterior angle
Intelligence
Size of interior angle
Special ability
Lines of symmetry
Agility
Order of rotational symmetry

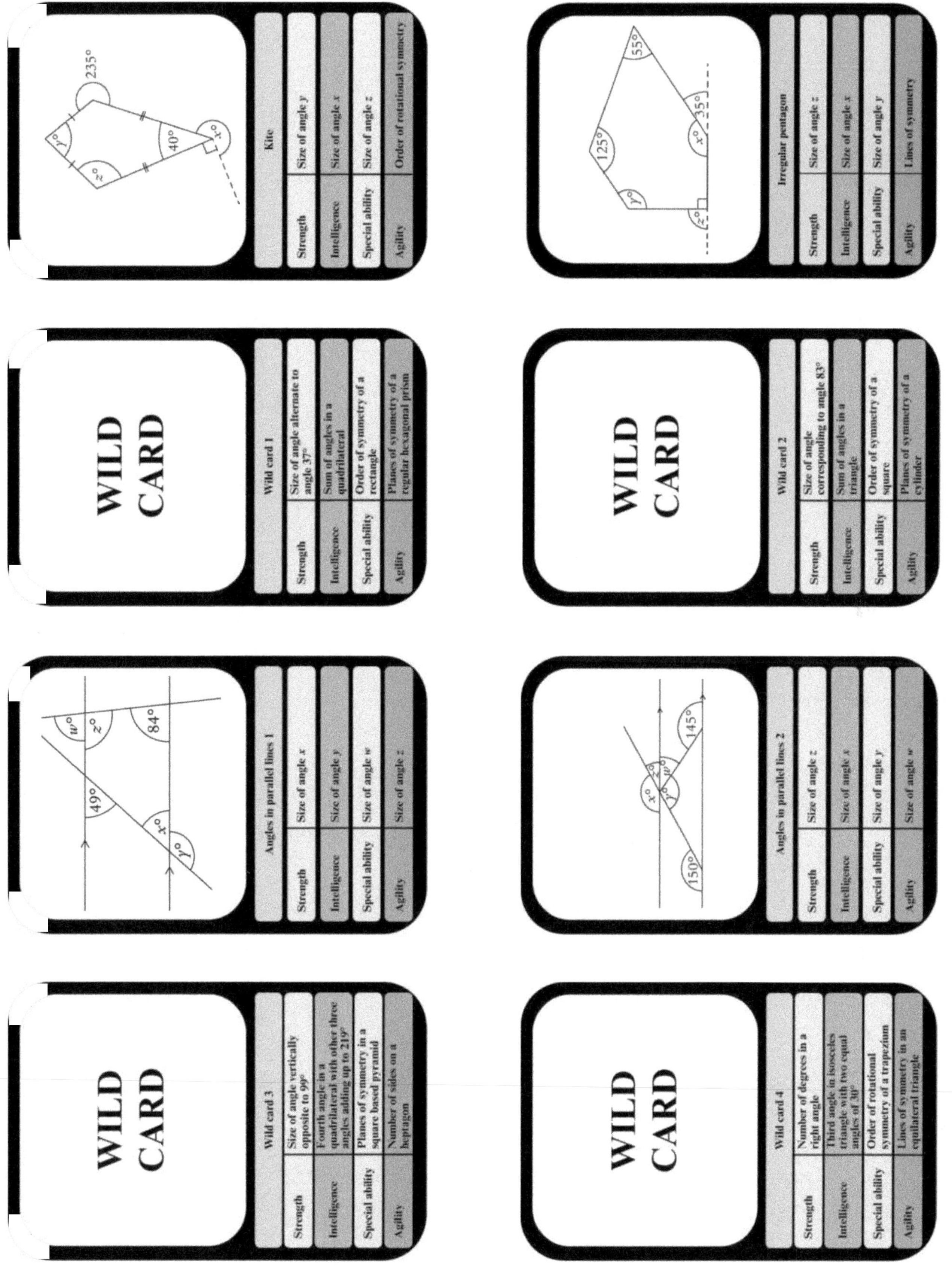

Larger size trump cards are available in the online resources.

Answers

Regular hexagon

S: 60

I: 120

SA: 6

A: 6

Regular pentagon

S: 72

I: 108

SA: 5

A: 5

Regular octagon

S: 45

I: 135

SA: 8

A: 8

Regular decagon

S: 36

I: 144

SA: 10

A: 10

Triangular prism

S: 60

I: 5

SA: 4

A: 6

Cube

S: 90

I: 6

SA: 5

A: 8

Parallelogram

S: 55

I: 125

SA: 55

A: 2

Isosceles trapezium

S: 40

I: 140

SA: 140

A: 1

Kite

S: 70

I: 230

SA: 125

A: 0

Irregular pentagon

S: 90

I: 145

SA: 125

A: 0

Wild card 1

S: 37

I: 360

SA: 2

A: 7

Wild card 2

S: 83

I: 180

SA: 4

A: ∞

Angles in parallel lines 1

S: 49

I: 131

SA: 84

A: 96

Angles in parallel lines 2

S: 30

I: 150

SA: 115

A: 35

Wild card 3

S: 99

I: 141

SA: 4

A: 7

Wild card 4

S: 90

I: 120

SA: 0

A: 3

The online resources contain blank top trump cards for students to make their own cards.

Ratio puzzle

Teacher notes

There are many errors made in ratio questions by misinterpreting the numbers given. The number given can be the total, the value for one quantity or occasionally the difference between one value and another, as in this puzzle. Students sometimes misread a question and so apply an incorrect method. This puzzle helps eliminate these misconceptions by introducing a ratio question with three plausible solutions, two of which are incorrect. This is intended as a revision or plenary exercise at the end of the ratio section in the student book.

Introductory activity

This activity should be completed at the end of the ratio chapter or as a revision activity later on in the course so little introduction should be necessary.

Puzzle activity

Students can work individually, in pairs or in groups to complete this activity in class, or it could be used as a homework activity. Photocopy the ratio problem sheet, which shows a ratio problem which has been solved using three different methods. Students are asked to decide which of the three methods is correct. For the two incorrect methods students need to think carefully what the error or misconception is. One possible way to do this is to re-write the question to match each answer.

Differentiation

To make this task easier, simply ask students to identify which is the correct method.

To make the task harder still see the alternative approach.

Alternative approaches

Ask students to use the blank template page, from the online resources, to be creative and write their own ratio puzzle. They should write the question and three attempts at the answer; two of which are wrong and one that is correct. This is quite a challenging task, and can be made harder by requiring all three attempts to be realistic whole numbers, which really requires some thought!

Answers

Attempt 1 Wrong

Possible question:

Ahraf, Ben and Carl share \$840 in the ratio 3 : 5 : 7.
How much does Ahraf get?

Attempt 2 Wrong

Possible question:

Ahraf, Ben and Carl share some money in the ratio 3 : 5 : 7. Carl gets \$840.
How much does Ahraf get?

Attempt 3 Correct

Ratio puzzle

Ahraf, Ben and Carl share some money in the ratio 3: 5: 7.
Carl gets \$840 more than Ben.
How much does Ahraf get?

Anne had three attempts at this question. Look at all three attempts below. Decide which is the correct method. For the methods that are incorrect re-write the question so that the method is then correct.

Attempt 1

$3 + 5 + 7 = 15$

$840 \div 15 = 56$

$3 \times 56 = 168$

Ahraf gets \$168

This method is correct ☐ wrong ☐ (please tick one box)

If you ticked wrong complete the following.

The question Anne is answering is:

Attempt 2

$840 \div 7 = 120$

$3 \times 120 = 360$

Ahraf gets \$360

This method is correct ☐ wrong ☐ (please tick one box)

If you ticked wrong complete the following.

The question Anne is answering is:

Attempt 3

$7 - 5 = 2$

$840 \div 2 = 420$

$3 \times 420 = 1260$

Ahraf gets \$1260

This method is correct ☐ wrong ☐ (please tick one box)

If you ticked wrong complete the following.

The question Anne is answering is:

Positive, negative and zero indices jigsaw

Teacher notes

This jigsaw activity is an interesting group work alternative to a written exercise. The advantages of a jigsaw activity over a written exercise are that it creates opportunities for students to discuss their ideas, and they can find an alternative starting point if they are not sure how to do a particular question. Students are able to start with the answer and work backwards, which is sometimes an interesting way to tackle a question.

This activity covers the laws of indices.

$x^a \times x^b = x^{a+b}$ $x^a \div x^b = x^{a-b}$

$(x^a)^b = x^{ab}$ $x^0 = 1$

$$x^{-a} = \frac{1}{x^a}$$

The activity uses positive, negative and zero indices but no fractional indices. Base numbers may be fractions.

Introductory activity

If you want to use this as a revision activity or a plenary you may find that little introduction is necessary. Otherwise you may want to revisit the laws of indices above, and you may need to revise reciprocals as there are questions requiring students to understand:

$$\left(\frac{x}{y}\right)^{-a} = \left(\frac{y}{x}\right)^{a}$$

Jigsaw activity

Ask your students to work in small groups. Copy the jigsaw pieces onto card to make a complete set for each group. There are several different-shaped jigsaws used in the Teacher Resource Kit so you should not tell the students what shape to expect. This jigsaw does not have any blank edges. This means that it is harder for your students to complete.

Differentiation

To make it more difficult, there are harder questions or 'solutions' including some answers that are common misconceptions around the outside edges of the jigsaw. If you want to make it easier you could indicate outside edges; perhaps by drawing a bold line along the outside edges.

Alternative approaches

You could just ask students to complete the jigsaw or you could make use of the fact that the outside edges are not blank. You could copy blank jigsaw pieces, from the online resources, and ask students to use these pieces to extend the jigsaw; either finding the solutions to the harder questions or writing a question that has the given solution. Alternatively students could make an entirely new jigsaw of their own.

Answers

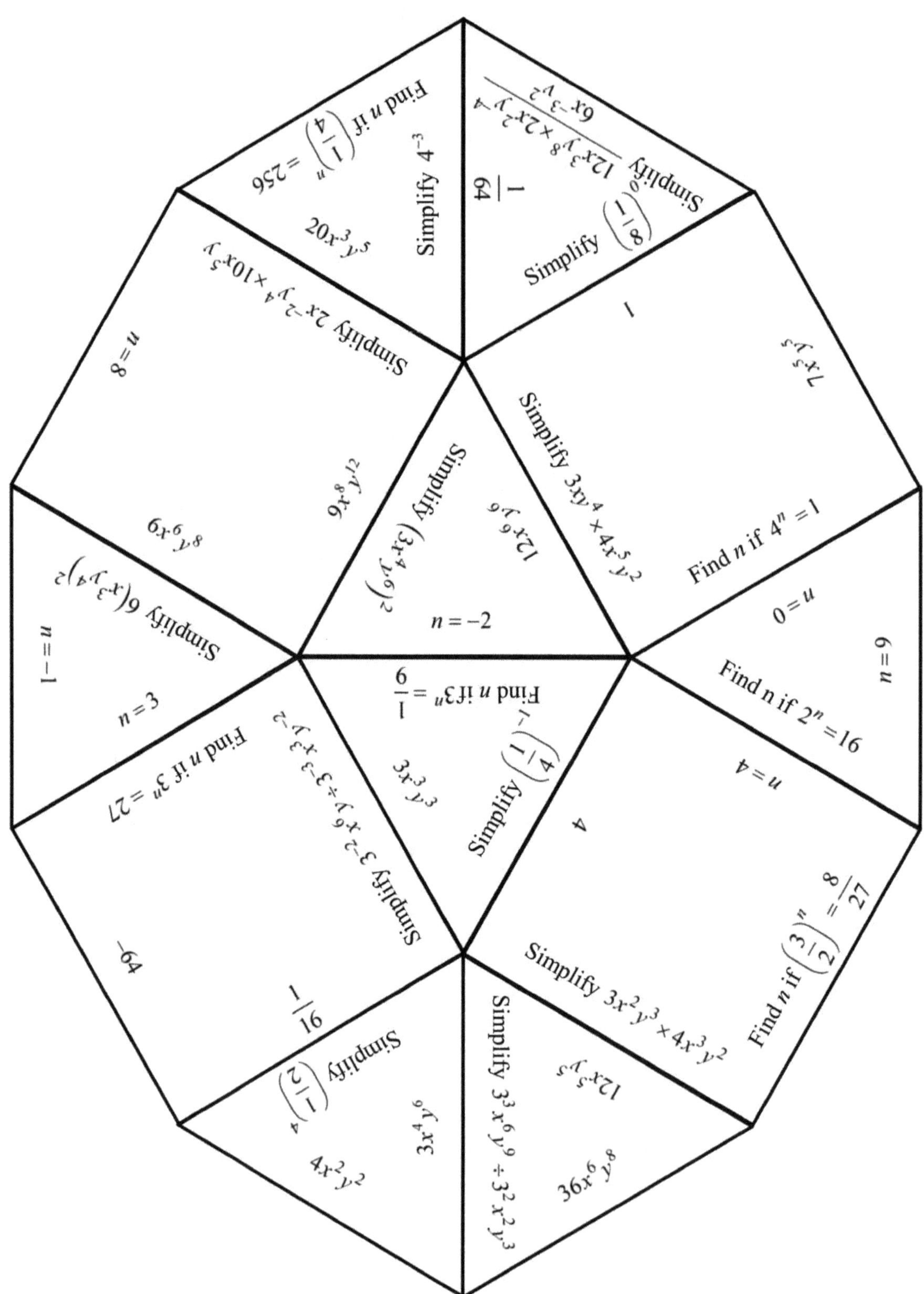

Find n if $\left(\frac{1}{4}\right)^n = 256$
Simplify 4^{-3}
$20x^3y^5$
Simplify $\frac{12x^3y^8 \times 2x^2y^{-4}}{6x^{-3}y^2}$
$\frac{1}{64}$
Simplify $\left(\frac{1}{8}\right)^0$
1
$7x^5y^5$
Simplify $2x^{-2}y^4 \times 10x^5y$
n = 8
$9x^8y^{12}$
$6x^9y^8$
Simplify $(3x^4y^6)^2$
$12x^6y^6$
n = −2
Simplify $3xy^4 \times 4x^5y^2$
Find n if $4^n = 1$
n = 0
n = 9
Simplify $6(x^3y^4)^2$
n = −1
n = 3
Find n if $3^n = \frac{1}{9}$
Find n if $2^n = 16$
n = 4
Find n if $3^n = 27$
Simplify $3^{-2}x^6y \times 3^{-3}x^3y^{-2}$
$3x^3y^3$
Simplify $\left(\frac{1}{4}\right)^{-1}$
4
Find n if $\left(\frac{3}{2}\right)^n = \frac{8}{27}$
−64
Simplify $3x^2y^3 \times 4x^3y^2$
$\frac{1}{16}$
Simplify $\left(\frac{1}{2}\right)^4$
$12x^5y^5$
Simplify $3^3x^6y^9 \div 3^2x^2y^3$
$3x^4y^6$
$4x^2y^2$
$36x^6y^8$

Positive, negative and zero indices jigsaw pieces

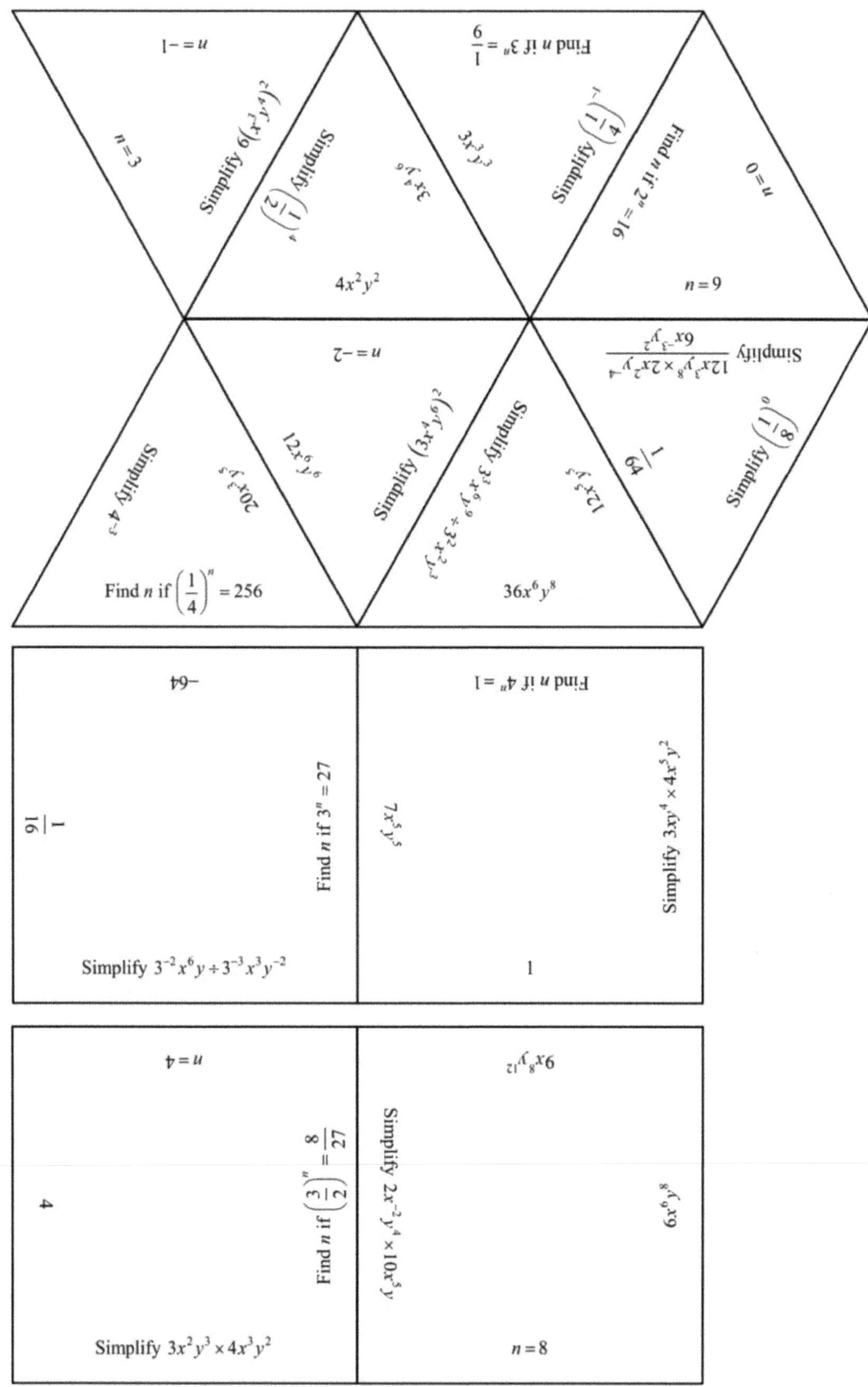

Larger size jigsaw pieces are available in the online resources.

Linear and non-linear graphs card sort

Teacher notes

One of the hardest concepts in the study of linear graphs is that there are many different ways you can write the equation of a straight-line (linear) graph. Students need to be able to identify the difference between linear and non-linear graphs. They also need to be able to identify key features of linear functions, such as the gradient and intercept. The purpose of the linear and non-linear graphs card sort activity is to give students the opportunity to compare and contrast linear and non-linear equations, so they can recognise the many different ways equations are presented.

Introductory activity

You could begin with an introductory activity. Write the following lists on the board and ask students: "What is the same? and "What is different….?" about the equations in each list

Linear equations: $y = 2y - 3$, $y = 5$, $y + x = 9$, $x = -10$, $3x - 2y = 4$

Non-linear equations: $y = x^2$ $y = 5x^3 - 2$ $y = \dfrac{1}{x}$ $x^2 + y^2 = 25$ $xy = 200$

Then give students two more equations, for example $x = 12$ and $y + x^2 = 3$, and ask them to decide which lists these should be in. This activity will help to clarify the difference between linear and non-linear graphs. To be able to work with gradients and intercepts your students will need some practice on rearranging equations into the form $y = mx + c$ where m is the gradient and c is the intercept. The most common misconception is that $y - 3x = 4$ has a negative gradient because there is a -3 in front of the x. Students need to rearrange to make y the subject to get $y = 3x + 4$ to see that the gradient is positive.

Card sorting activity

Once a few rules have been established on what makes a linear or non-linear graph, and students have been reminded how to rearrange equations into the form $y = mx + c$, then move on to the card sorting activity. Copy sets of cards and cut them up, ask students to work in small groups. You could simply ask students to sort the cards into two piles: linear graphs and non-linear graphs.

Differentiation

You can allow students a "**don't know**" pile. If you want to extend the activity ask the students to think about the gradients of the linear equations and sort the linear pile into four further categories: those with: **positive gradients**, those with **negative gradients**, **horizontal lines** and **vertical lines**. Alternatively think about the y-intercepts of the lines and ask students to sort the linear cards into four categories, those that: **pass through the origin**, have a **positive *y*-intercept**, have a **negative *y*-intercept** or have **no *y*-intercept**.

You can also differentiate by removing harder cards for less able students or by adding harder cards to extend the more able students. The cards can be adjusted as you move around the class.

Alternative approaches

To allow students a bit more freedom, simply ask them to sort the cards, leaving them to choose the categories. You could then ask students to add some further examples of their own. The students then have a choice to find an easy example or a really difficult one. You could give them the incentive to try a difficult one by telling them their card will be given to another group for the other group to try to classify it. You can also include equations without x and y in them, for example conversion graphs or distance–time graphs.

Answers

Linear				Non-linear
Horizontal lines	**Vertical lines**	**Positive gradient**	**Negative gradient**	
$y = -2$	$x = 0$	$y = 3x$	$y = 6 - 5x$	$x^2 + y^2 = 16$
$y = 5$	$x = \frac{1}{4}$	$y = 4x + 1$	$y - 10 = -x$	$xy = 9$
$0 = y$	$3 = x$	$y = \frac{5x}{3}$	$y + 5x = 4$	$x = \frac{1}{y}$
$y = 0.4$	$x = -1$	$y - 4 = 3x$	$2y - 5 = -7x$	$x = y^2 - 2$
		$x = y$	$6y - 8 + 3x = 0$	$x^2 = y + 4$
		$y + 1 = 4x$	$4x = 12 - y$	$y = x^3 + 5x^2 - 4x$
		$3y - 2x = 10$	$x + y = 0$	$y = \frac{5}{x}$
		$x = 5 + y$	$y = -\frac{x}{10}$	
		$y + 8 - 5x = 0$		
Passes through origin	**On y-axis intercept**	**Positive intercept**	**Negative intercept**	
$0 = y$	$x = \frac{1}{4}$	$y = 5$	$y = -2$	
$x = 0$	$3 = x$	$y = 0.4$	$y + 1 = 4x$	
$y = 3x$	$x = -1$	$y = 4x + 1$	$x = 5 + y$	
$y = \frac{5x}{3}$		$y = 6 - 5x$	$y + 8 - 5x = 0$	
$y = -\frac{x}{10}$		$y - 10 = -x$		
$x + y = 0$		$y - 4 = 3x$		
$x = y$		$y + 5x = 4$		
		$3y - 2x = 10$		
		$2y - 5 = -7x$		
		$6y - 8 + 3x = 0$		
		$4x = 12 - y$		

Alternative extra equations you could include that are linear but don't involve x and y.

$k = m$ $\quad$ $F = 1.8C + 32$ $\quad$ $c = 2.5i$

$m = 7 + 3p$ $\quad$ $r = t$ $\quad$ $s = 3t$

Linear and non-linear graphs sort cards

$y = -2$	$x^2 + y^2 = 16$	$4x = 12 - y$
$y = 3x$	$y = 4x + 1$	$x + y = 0$
$y - 10 = -x$	$y = x^3 + 5x^2 - 4x$	$3y - 2x = 10$
$xy = 9$	$0 = y$	$3 = x$
$y = \frac{5x}{3}$	$x = 0$	$y = -\frac{x}{10}$
$y - 4 = 3x$	$x = -1$	$y + 1 = 4x$
$y = 5$	$x = 5 + y$	$y = \frac{5}{x}$
$y + 5x = 4$	$y = 6 - 5x$	$x = y$
$x^2 = y + 4$	$x = y^2 - 2$	$y = 0.4$
$2y - 5 = -7x$	$6y - 8 + 3x = 0$	$y + 8 - 5x = 0$
$x = \frac{1}{y}$	$x = \frac{1}{4}$	

Larger size sort cards are available in the online resources.

Arc length and sector area treasure hunt activity

Teacher notes

This treasure hunt activity is an interesting and enjoyable alternative to a written exercise allowing students to discuss their ideas and move around the classroom. If they are not sure how to do a particular question students can find an alternative starting point as the activity works no matter where the students begin. This activity requires students to be able to work with questions involving arc length and sector area. There are also some questions where students are given the arc length or sector area and are asked to work backwards to find either the radius or the unknown angle, θ. Students also need to be confident at rounding to three significant figures.

Introductory activity

If you want to use this as a revision activity or a plenary you may find that little introduction is necessary, if not you may want to go through these two basic examples.

Example 1

Find the area of this sector.

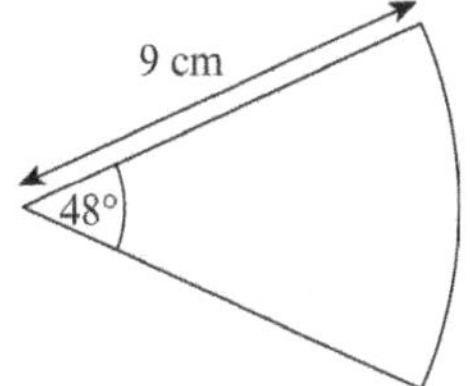

$$\text{area} = \frac{48}{360} \times \pi \times 9^2$$
$$= 33.9 \text{ cm}^2$$

Example 2

Find the arc length of this sector.

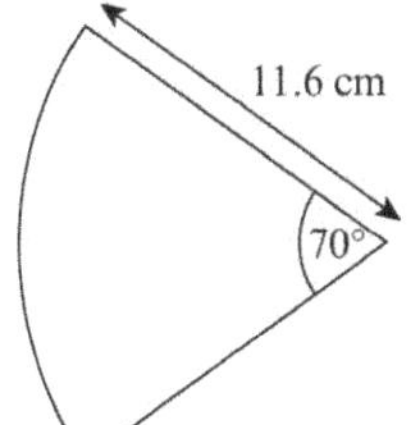

$$\text{arc length} = \frac{70}{360} \times 2 \times \pi \times 11.6$$
$$= 14.2 \text{ cm}$$

Treasure hunt activity

Ask your students to work in small groups. Tell them that all answers need to be given to 3 significant figures. Each group requires a copy of the blank answer sheet to record their answers. Copy the game cards onto card and cut them up (larger size game cards are available in the online resources). Stick the individual cards randomly around the classroom on the walls. Each card contains a card number at the top. This is both the card number and the answer to the previous question. Students pick any card to be their starting point, they need to write the card number anywhere on their answer grid. Then they answer the question on the bottom section of the card and look for that answer in the classroom. This will be the next number in the loop that they write down on their answer grid. Students will know when they have finished the activity because their answer sheet will have a number in every section.

Differentiation

To make it harder you could replace the bottom part of one or two cards with the words "You need to write a question related to circles with the answer …" and give them the answer to the next card. Or students can use the blank treasure hunt cards, from the online resources, to make an entirely new treasure hunt activity of their own. Blank answer sheets are also provided.

Alternative approaches

This does not have to be done as a treasure hunt activity, if you prefer the students not to move around the classroom, instead sets of cards can be placed on desks and the activity can still be done in the same way.

Previous answer:

50.3 cm

Question:

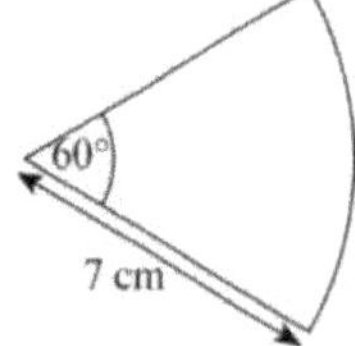

For this sector work out the arc length.

Next answer?

Previous answer:

7.33 cm

Question:

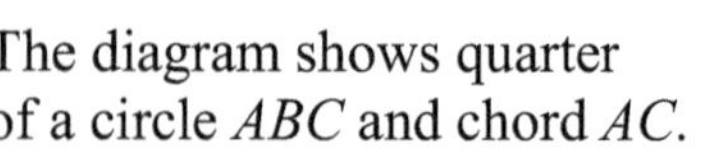

The diagram shows quarter of a circle ABC and chord AC.

17.5 cm

A B C

Calculate the area of the shaded region.

Next answer?

Previous answer:

87.4 cm^2

Question:

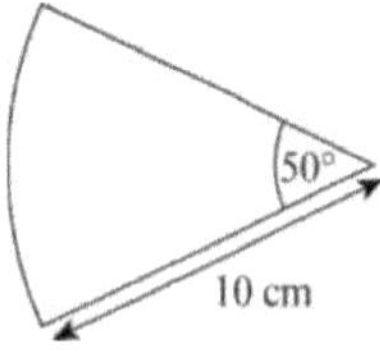

For this sector work out the perimeter.

Next answer?

Previous answer:

28.7 cm

Question:

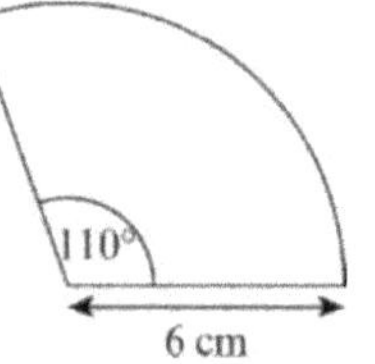

For this sector work out the sector area.

Next answer?

Previous answer:

34.6 cm^2

Question:

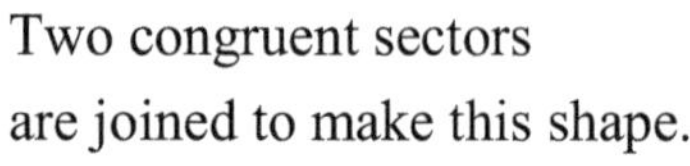

Two congruent sectors are joined to make this shape.

Find the perimeter of the shape.

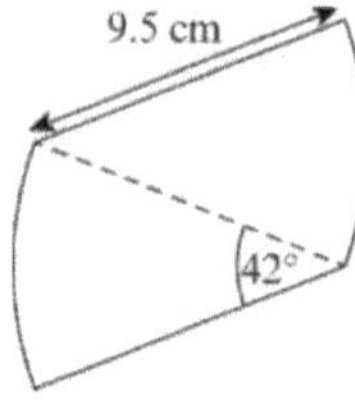

Next answer?

Previous answer:

32.9 cm

Question:

The curved parts of the shaded shape are both arcs from the same centre.

5 cm

50°

3 cm

Work out the area of the shaded shape.

Next answer?

Previous answer:

17.5 cm^2

Question:

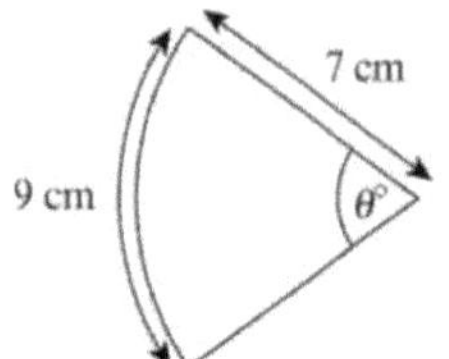

For this sector work out the size of the angle θ.

Next answer?

Previous answer:

73.6°

Question:

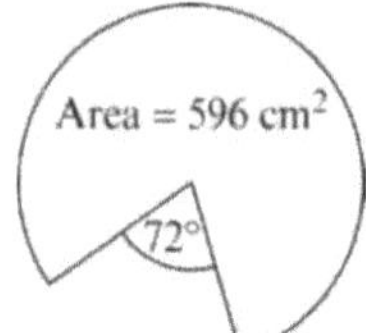

For this sector work out the radius.

Next answer?

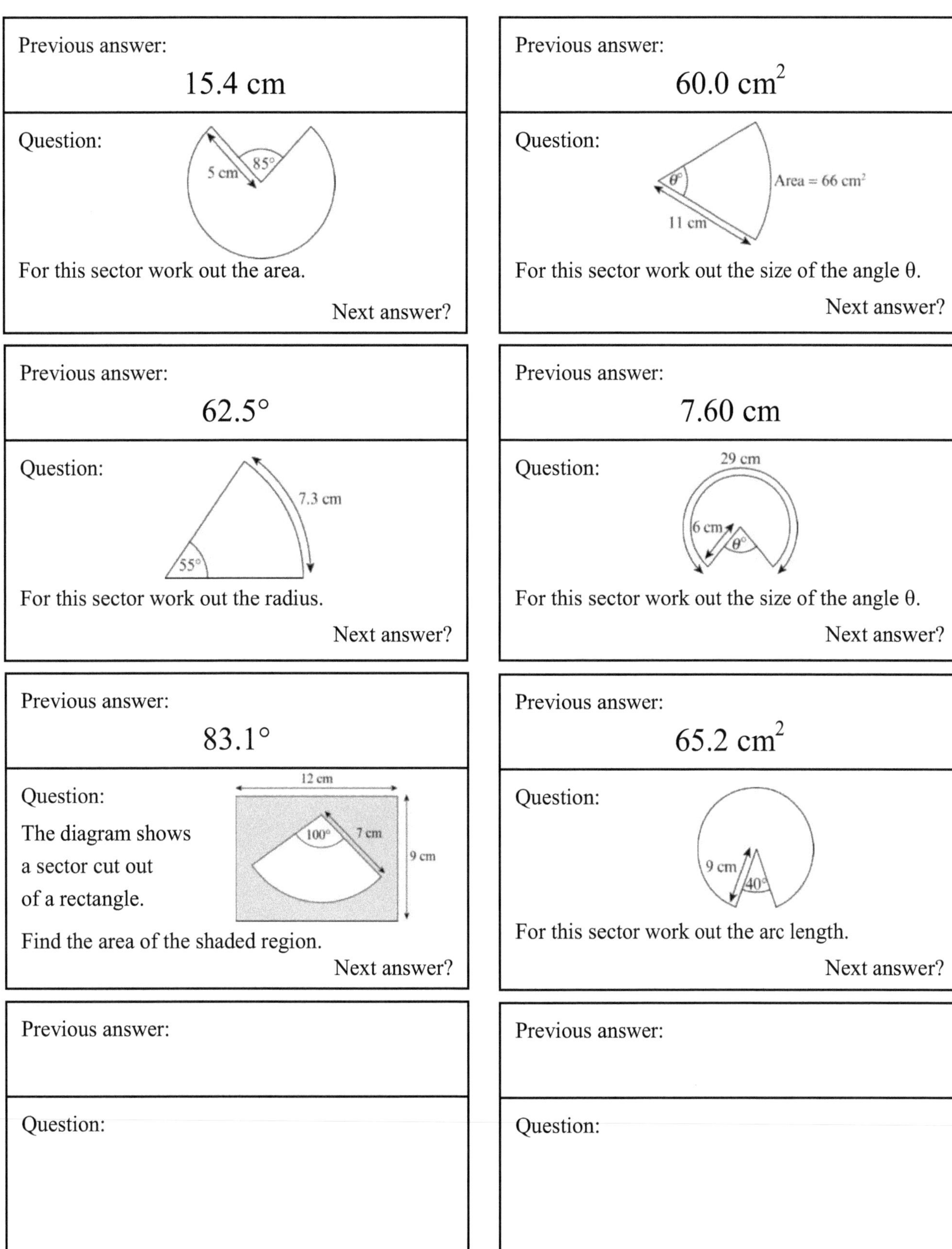

Previous answer:

15.4 cm

Question:

For this sector work out the area.

Next answer?

Previous answer:

60.0 cm^2

Question:

For this sector work out the size of the angle θ.

Next answer?

Previous answer:

62.5°

Question:

For this sector work out the radius.

Next answer?

Previous answer:

7.60 cm

Question:

For this sector work out the size of the angle θ.

Next answer?

Previous answer:

83.1°

Question:

The diagram shows a sector cut out of a rectangle.

Find the area of the shaded region.

Next answer?

Previous answer:

65.2 cm^2

Question:

For this sector work out the arc length.

Next answer?

Previous answer:

Question:

Next answer?

Previous answer:

Question:

Next answer?

Further blank cards, and blank answer sheets, are available in the online resources.

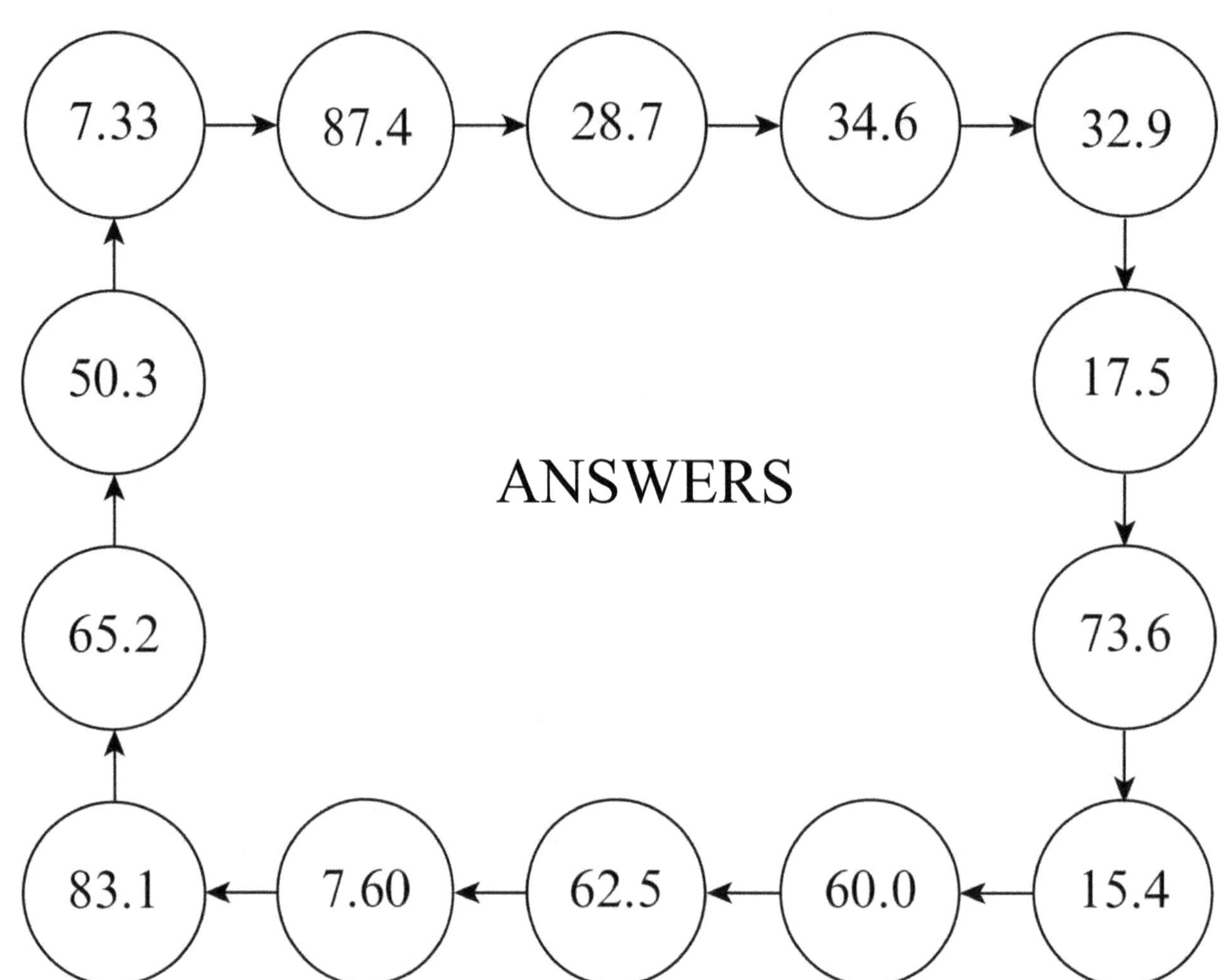
7.33
87.4
28.7
34.6
32.9
17.5
ANSWERS
73.6
15.4
60.0
62.5
7.60
83.1
65.2
50.3

Averages, range and bar charts card sort

Teacher notes

Students are usually competent calculating the three averages, mean, median and mode, when the data is presented as a list of numbers or in a frequency table. However problems can arise when students are asked to extract information from data in the form of a bar chart. The purpose of this card sort activity is to allow students to compare and contrast the shapes of different bar charts so that they can interpret them and their relationships with the three averages and the range. Note all of these charts have 15 data values, you could do some extra sets with more or less values.

Introductory activity

You may want to use this activity as a revision activity in which case you may choose to give little or no introduction.

Alternatively you may want to remind students of the three different averages and the range.

- The mode is the value which occurs the most.
- The median is the value in the middle when all the data is ordered.
- The mean is the total of all data values divided by the number of data values.
- The range is the highest data value minus the lowest data value.

Card sorting activity

Print out the cards from the online resources and cut them up. Ask students to work in small groups. Give each group a set of these cards. You may want to use different coloured card for each set so that if any are dropped it is easy to see which set they belong to. To keep it simple you could ask students to sort the cards into matching pairs. If you want to extend the activity ask the students to write down the strategies they used for sorting the cards, or ask students to order the cards by ease of working out. They will probably say mode is easiest as you can see it straight away in a bar chart, followed by range, etc.

Differentiation

You can differentiate by leaving some difficult cards out for less able students or by adding harder cards to extend the more able students. You can allow students a "don't know" pile. These can be sorted when you move around the class.

Alternative approaches

You could extend the task further, by asking students to write some examples of their own. The students then have a choice to find easy examples or really difficult ones. You could give them the incentive to do a difficult one by telling them their card will be given to another group for the other group to try to classify it. You could use the blank cards from the online resources or you could blank out one or more of the numbers in some or all of the tables and then ask students to fill the blanks.

Answers

The cards are provided in the correct pairs and will therefore need shuffling before use.

Averages, range and bar chart sort cards

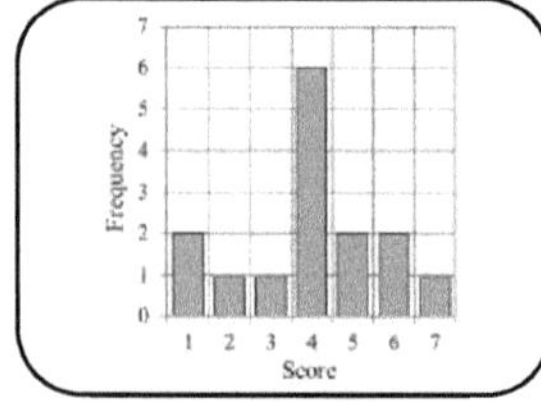

mean	4
median	4
mode	4
range	6

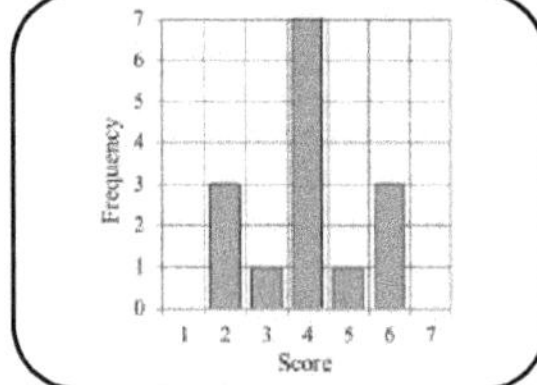

mean	4
median	4
mode	4
range	4

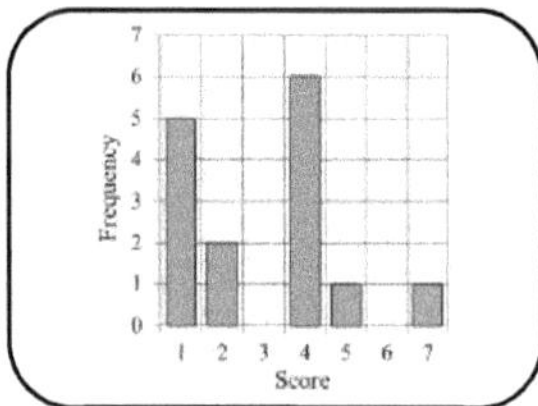

mean	3
median	4
mode	4
range	6

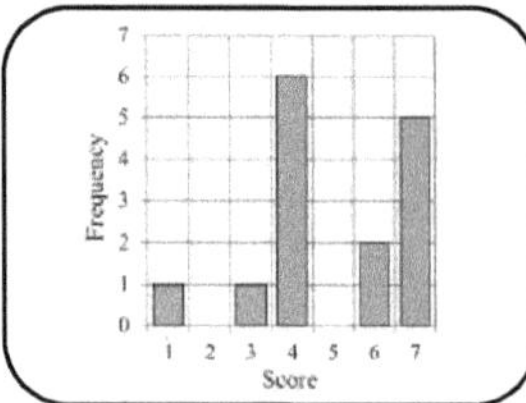

mean	5
median	4
mode	4
range	6

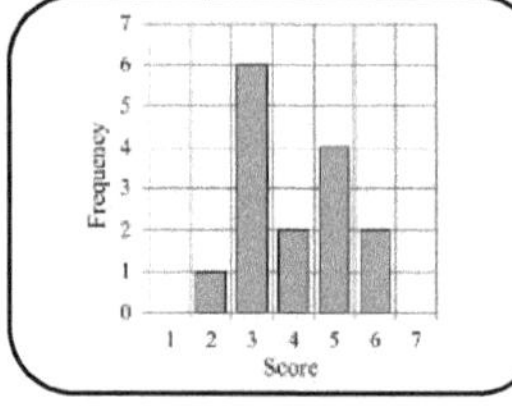

mean	4
median	4
mode	3
range	4

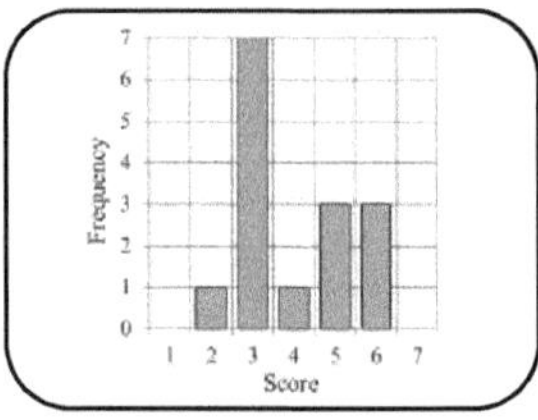

mean	4
median	3
mode	3
range	4

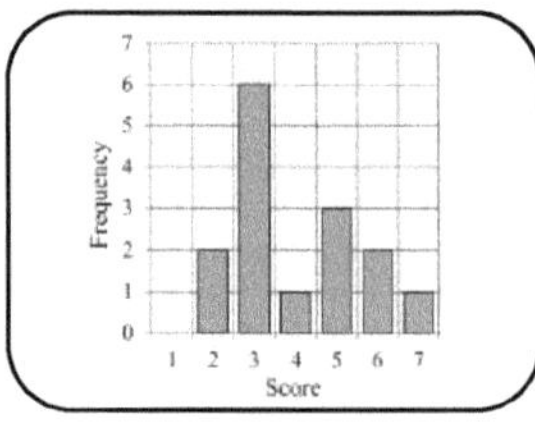

mean	4
median	3
mode	3
range	5

The cards are given in the correct pairs.

Larger size sort cards are available in the online resources.

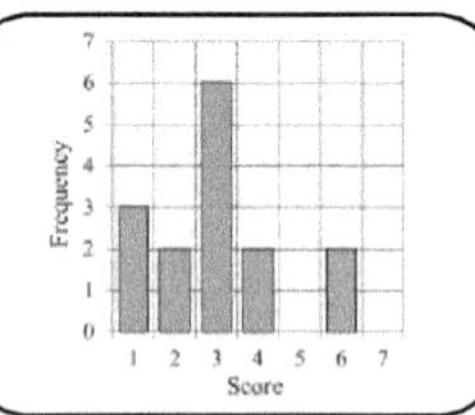

mean	3
median	3
mode	3
range	5

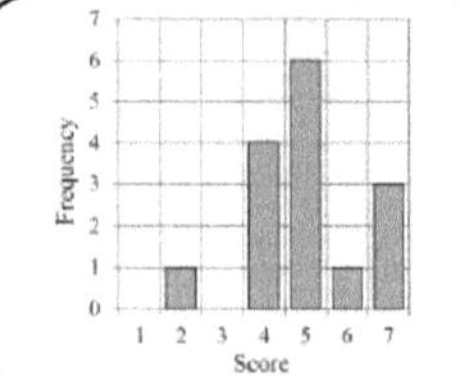

mean	5
median	5
mode	5
range	5

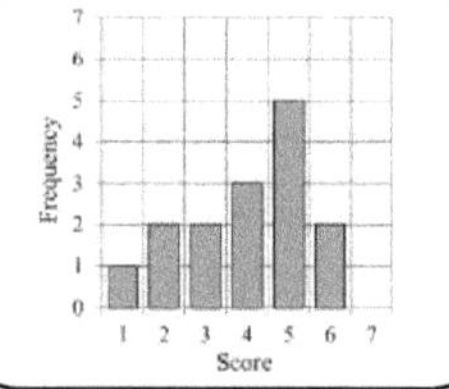

mean	4
median	4
mode	5
range	5

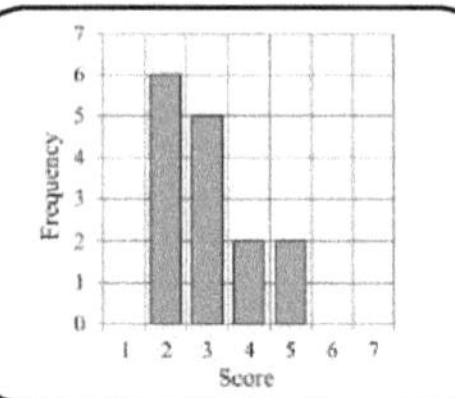

mean	3
median	3
mode	2
range	3

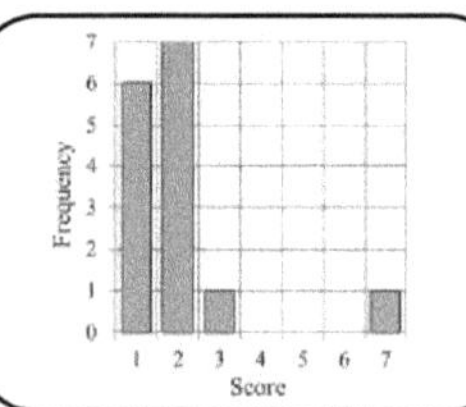

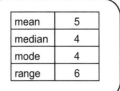

mean	2
median	2
mode	2
range	6

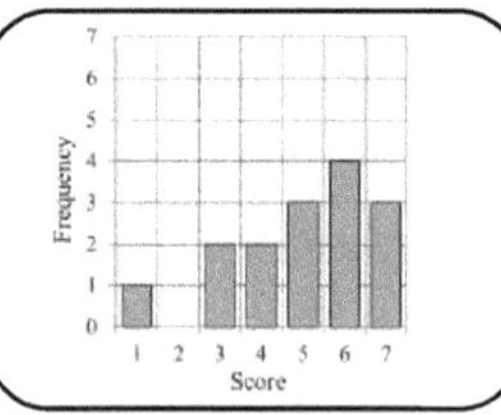

mean	5
median	5
mode	6
range	6

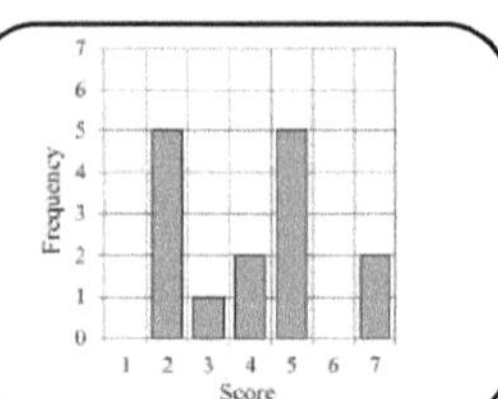

mean	4
median	4
mode	2 & 5
range	5

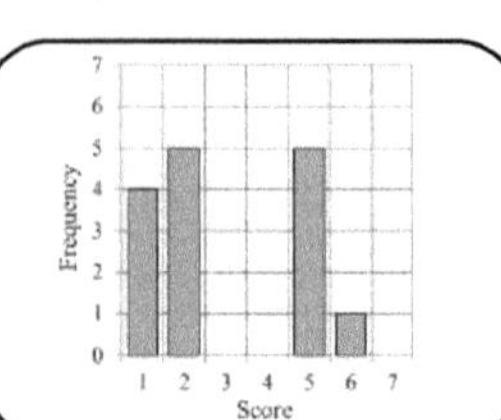

mean	3
median	2
mode	2 & 5
range	5

Standard form jigsaw

Teacher notes

This jigsaw activity is an interesting group work alternative to a written exercise. The advantages of a jigsaw activity over a written exercise are that it creates opportunities for students to discuss their ideas, and they can find an alternative starting point if they are not sure how to do a particular question. Students are able to start with the answer and work backwards, which is sometimes an interesting way to tackle a question. This activity requires students to be able to work with standard form, to be able to convert very large and very small numbers into standard form, and to be able to convert numbers in index form into standard index form.

Introductory activity

To use this as a revision activity or a plenary, you may find that little introduction is necessary. Otherwise you may want to go through the following examples.

Example 1

28 300 000 in standard form is 2.83×10^7

Note a common error is to count the zeros and write 2.83×10^5 this task will go some way to addressing this.

Example 2

0.000045 in standard form is 4.5×10^{-5}

Example 3

412.5×10^3 is not in standard form because 412.5 is not between 1 and 10, but it can be converted to standard form as follows.

Step 1 Write 412.5 in standard form 4.125×10^2

Step 2 Rewrite 412.5×10^3 as $4.125 \times 10^2 \times 10^3$
$= 4.125 \times 10^5$

Example 4

0.032×10^{-5} is not in standard form because 0.032 is not between 1 and 10, but it can be converted to standard form as follows.

Step 1 Write 0.032 in standard form 3.2×10^{-2}

Step 2 Re-write 0.032×10^{-5} as $3.2 \times 10^{-2} \times 10^{-5}$
$= 3.2 \times 10^{-7}$

Example 5

If $432\ 000\ 000 = 4.32 \times 10^n$ what is the value of n? Converting to standard form gives us $n = 8$.

Jigsaw activity

Ask your students to work in small groups. Copy the jigsaw pieces onto card to make a complete set for each group. There are several different-shaped jigsaws in the Teacher Resource Kit so you should not tell the students what shape to expect. This jigsaw has blank edges so it is easier for your students to complete.

Differentiation

To make the task easier you could simply ask students to do the jigsaw, to make it harder you could remove one or two of the pieces from the middle giving them blank pieces, and asking them to complete these pieces to fill in the gaps. Blank pieces can be found in the online resources.

Alternative approaches

You could copy blank jigsaw pieces from the online resources and ask students to use these pieces to make an entirely new jigsaw of their own, or ask them to consider extending this jigsaw. Note, if you intend asking them to extend this jigsaw students will need to write on the blank edges of the printed jigsaw.

Answers

0.46×10^7 $n = -4$	This is not in standard form 0.0000641 in standard form is Write three million in standard form	3×10^6 Subtract the indices
If $0.000518 = 5.18 \times 10^n$ what is the value of n? If $32\,140\,000 = 3.214 \times 10^n$ what is the value of n? Add the indices	6.41×10^{-5} $n = 7$ 82 000 000 in standard form is 820×10^4 is not in standard form change it to standard form	To find $10^6 \div 10^2$ 8.2×10^6 3×10^5
To find $10^7 \times 10^3$ $n = -5$ 8.2×10^{-2}	8.2×10^7 If $0.0000263 = 2.63 \times 10^n$ what is the value of n? 641 000 in standard form is Standard index form	Write the answer to 1500×200 in standard form Standard form is also known as 8.2×10^{-4}
0.082 in standard form is 6.41×10^7	6.41×10^5 64 100 000 in standard form is $n = 4$	0.82×10^{-3} is not in standard form change it to standard form If $47\,500 = 4.75 \times 10^n$ what is the value of n?

Standard form jigsaw pieces

Subtract the indices 3×10^6	To find $10^6 \div 10^2$ 8.2×10^6 3×10^5	8.2×10^{-4} Standard form is also known as Write the answer to 1500×200 in standard form
6.41×10^{-5} $n = 7$ 820×10^4 is not in standard form change it to standard form 82 000 000 in standard form is	$n = -4$ 0.46×10^7	6.41×10^7 0.082 in standard form is

Add the indices If $0.000518 = 5.18 \times 10^n$ what is the value of n? If $32\,140\,000 = 3214 \times 10^n$ what is the value of n?	0.82×10^{-3} is not in standard form change it to standard form If $47\,500 = 4.75 \times 10^n$ what is the value of n?	This is not in standard form 0.0000641 in standard form is Write three million in standard form
6.41×10^5 64 100 000 in standard form is $n = 4$	8.2×10^{-2} $n = -5$ To find $10^7 \times 10^3$	641 000 in standard form is Standard index form If $0.0000263 = 2.63 \times 10^n$ what is the value of n? 8.2×10^7

Larger size jigsaw pieces are available in the online resources.

Simultaneous equations correct my homework activity

Teacher notes

We often ask our students to make sure that they have checked their working but this is a task that we rarely teach them how to do. The purpose of this activity is to help students understand what they should focus on when checking their working in simultaneous equations questions.

Students sometimes make careless errors such as reading points incorrectly from the graph (as in the solution to the first question in this task). One of the biggest problems in algebraic work is dealing correctly with negatives. Students often miss out the final step of substituting both values into both the equations to check. This is their opportunity to "mark their own work" yet it is frequently not seen! To do this task students should be familiar with two of the methods for solving simultaneous equations: graphically and by elimination.

Introductory activity

Too much introduction and explanation will make this task easy and may well spoil it for the students. The main errors in the working are arithmetic; for example, not dealing correctly with the negatives or not multiplying every term in an equation. There is also a reading off a graph error.

This task may follow on from the work on simultaneous equations in the student book, or be used as a revision activity later in the course.

A quick starter on negative numbers, using five quick questions to check the basics, will help ensure that there are no negative number misconceptions before beginning the task.

Question		Answer
1	$-3 \times -5 =$	15
2	$-4 \times 7 =$	-28
3	$-2 - 5 =$	-7
4	$8 - (-4) =$	$8 + 4 = 12$
5	$x = 4$, $y = -3$ what is $2x - 4y$?	$2 \times (4) - 4 \times (-3)$ $= 8 + 12 = 20$

Correct my homework activity

Ask your students to work in small groups or pairs for this activity so that they can discuss ideas. Give each group, or pair, a copy of the "Correct my homework" activity sheet. Ask students to use the right-hand column in the table to write down an explanation for the working that has been done and to make any necessary corrections.

Differentiation

To make this task a little easier you can give the students a few of the answers in the second column. To make this task a bit harder miss out one or two lines of working in the first column. For example, miss out the line $2y = 2 - 3 \times (-2)$ from the last but one row in the table.

Alternative approaches

Ask students to solve their own question on simultaneous equations using either the graphical method or the elimination method (or any other method that they have learned). Ask students to make deliberate mistakes in the working and then give this incorrect solution to a different group to correct.

Answers

Q	Explanation/Corrections
1)	The student has decided to solve the first pair of simultaneous equations using … a graphical … method. They have begun by correctly drawing a table of results to help them draw the lines. The first table is for $y = x + 3$ and the second table is for $y = 5 - x$.
	The student has correctly drawn the two lines but has incorrectly read off the point of intersection. The correct answer should be: (1, 4) or $x = 1, y = 4$. In fact if you look in the two tables, in this case, you can actually see the solution without having to draw the graph.
2)	The student has correctly rearranged the first equation so that the algebraic terms are on the left hand side.
	The student has tried to multiply the first equation by 3 but has forgotten to multiply the 2 by 3. The correct first line of working here should be: $9x + 6y = 6$
	The student has made the correct decision to subtract the two equations however whilst the x and y terms on the left have been subtracted the number terms on the right have been added. The (follow through) working here should have been: $9x + 6y = 2$ / $-(\underline{5x + 6y = -10})$ / $\underline{4x \quad = 12}$ followed by $x = 3$ or using the **correct RHS**: $9x + 6y = 6$ / $-(\underline{5x + 6y = -10})$ / $\underline{4x \quad = 16}$ followed by $x = 4$
	This follows correctly from their last line of working where they have divided – 8 by 4, however the correct answer is actually $x = 16 \div 4 = 4$
	The student is substituting their value of $x = -2$ into a correct rearrangement of the first equation. However they have made a mistake as they have calculated $-3 \times -2 = -6$ instead of $+6$.This is in addition to their previous mistakes. The fully correct working is: $2y = 2 - 3x$ $2y = 2 - 3 \times (4) = 2 - 12 = -10$ $y = -5$ So the solution is $x = 4$ and $y = -5$ Check in $5x + 6y = -10$: $5 \times (4) + 6 \times (-5) = 20 - 30 = -10$ correct!
	The student has not checked their work, had they substituted both their values into the second equation (not previously used) they would have seen that they were incorrect. Substitute $x = -2$ and $y = -2$ into $5x + 6y$ $5 \times (-2) + 6 \times (-2) = -10 - 12 = -22$ and not the –10 it should be.

Simultaneous equations correct my homework

In the working below there are some deliberate mistakes. Your task is to explain what is happening in the working this student has shown. The first box has been started for you. You also need to correct any errors that you find.

Solve these simultaneous equations using a method of your choice.

1) $y = x + 3$ and $y = 5 - x$ **2)** $3x = 2 - 2y$ and $5x + 6y = -10$

Working

1)

x	–1	0	1	2
y	2	3	4	5

x	–1	0	1	2
y	6	5	4	3

Explanation/Corrections

The student has decided to solve the first pair of simultaneous equations using .... method.

....

....

....

....

(4,1) or $x = 4, y = 1$

---

2) $3x = 2 - 2y$

$5x + 6y = -10$

$3x + 2y = 2$

$5x + 6y = -10$

$9x + 6y = 2$

$5x + 6y = -10$

$$\begin{array}{rl} & 9x + 6y = 2 \\ - & 5x + 6y = -10 \\ & 4x \quad\;\; = -8 \end{array}$$

$x = -2$

$2y = 2 - 3x$

$2y = 2 - 3 \times (-2)$

$2y = 2 - 6$

$2y = -4$

$y = -2$

$x = -2$ and $y = -2$

The student has missed out an important piece of working at the end. What do you think this is?

A larger version is available in the online resources.

Transformation game

Teacher notes

One of the main problems when working with transformations is understanding how to fully describe a transformation. For example, if describing a rotation, students must state the angle, direction and centre of rotation; they will often forget one of these. Students can also find combinations of transformations difficult. This is an interesting, alternative approach to using a written exercise as it promotes discussion, allows for group work, and the students often find it enjoyable. This task creates opportunities for students to discuss their ideas, and if they are not sure how to do a particular transformation they can complete an alternative one to help them gain confidence. To maximise scores in the game they will need to work out all the different transformations and the context of a game gives an incentive to do this.

Introductory activity

You may want to do a trial game to explain the rules, or give students a photocopy of the game instructions, available in the online resources, and talk through these instructions, in particular the scoring system.

You could revise how to describe fully the four transformations.

- Rotation: the angle, the direction and the centre of rotation.
- Reflection: the equation of the mirror line.
- Translation: a direction vector.
- Enlargement: the scale factor and centre of enlargement.

You may also want to work through an example of how to combine transformations.

Game activity

Ask your students to work in small groups of two to four players. Each group will need:

- One copy of the transformation board; you may want to enlarge this when copying.
- A set of the transformation cards. You will need to copy these onto card and cut them up to make a total of 48 cards. You may want to use different coloured card for each group so that if a piece is dropped or lost you know which group it belongs to.
- A counter.
- Paper and a pen to record scores.
- A copy of game instructions and rules.

How to play

1. Put the counter on the start triangle marked with an S.
2. Shuffle the cards face down and deal four to each player.
3. Place the rest of the pack face down on the table in an "unused pile".
4. Players take it in turns to move; for each turn they must try to move the counter from its current position to another triangle on the board using one of their transformation cards. (See differentiation.)
5. When a player has moved, they must throw away the used card onto a "discard pile" and collect a new card from the unused pile.
6. If a player cannot move they may discard one of their cards and collect a new one from the unused pile.
7. The cards are graded according to difficulty. For each move using a card A scores 4 points, a move using a card B scores 3 points, a move using a card C scores 2 points and a move using a card D scores 1 point.

8 If using transformation cards B, C or D players must describe in full the transformation they are doing, if they fail to do so they do not score the points.
9 If a player makes a mistake, the counter returns to its previous position and the player misses a turn and loses their card.
10 The game ends when the unused pile is empty and no more moves are possible.
11 The player with the highest score wins.

Differentiation

To make the game a little easier deal five cards to each player instead of four. You could allow the use of tracing paper to help them with the transformations, or you could allow students to draw the mirror lines $y = x$ and $y = -x$ on the game board.

You could make it harder by allowing combinations of transformations (step **4**); two or more cards per turn could be used. If they use more than one card, you could decide that the intermediate positions of the transforming triangle do not need to be marked on the game board. The desire to score more points will encourage students to attempt complex combinations of transformations. To really challenge the students ask them to design a different scoring system, for example, some of the difficult-to-reach triangles on the board could be assigned additional bonus scores. This will encourage students to consider which triangles are easier/harder to get to.

Alternative approaches

You could give the students a copy of the blank game board and sort cards, from the online resources, and ask them to produce their own version of this game using different shapes and including some more transformations that are different transformations to the ones already included. For example, they could extend the allowed transformations to include stretches or shears. You could also combine this with work on matrix transformations and give bonus points for the correct matrix of each transformation. Matrices for those card A transformations that are possible are given below.

Answers

If you decide to extend this activity to the alternative approach and award bonus points for describing the matrices that match the enlargement, rotation and reflection transformations, the possible matrices are as follows.

Transformation	Matrix
Rotations	
Centre: origin; Angle: 90°; Direction: clockwise OR Centre: origin; Angle: 270°; Direction: anticlockwise	$\begin{pmatrix} 0 & 1 \\ -1 & 0 \end{pmatrix}$
Centre: origin; Angle: 90°; Direction: anticlockwise OR Centre: origin; Angle: 270°; Direction: clockwise	$\begin{pmatrix} 0 & -1 \\ 1 & 0 \end{pmatrix}$
Centre: origin; Angle: 180°; Direction: not applicable	$\begin{pmatrix} -1 & 0 \\ 0 & -1 \end{pmatrix}$
Reflections	
In the y-axis (the line $x = 0$)	$\begin{pmatrix} -1 & 0 \\ 0 & 1 \end{pmatrix}$
In the x-axis (the line $y = 0$)	$\begin{pmatrix} 1 & 0 \\ 0 & -1 \end{pmatrix}$
In the line $y = x$	$\begin{pmatrix} 0 & 1 \\ 1 & 0 \end{pmatrix}$
In the line $y = -x$	$\begin{pmatrix} 0 & -1 \\ -1 & 0 \end{pmatrix}$
Enlargements	
Centre: origin; Scale Factor: 2	$\begin{pmatrix} 2 & 0 \\ 0 & 2 \end{pmatrix}$
Centre: origin; Scale Factor: $\frac{1}{2}$	$\begin{pmatrix} \frac{1}{2} & 0 \\ 0 & \frac{1}{2} \end{pmatrix}$
Centre: origin; Scale Factor: –2	$\begin{pmatrix} -2 & 0 \\ 0 & -2 \end{pmatrix}$
Centre: origin; Scale Factor: –1	$\begin{pmatrix} -1 & 0 \\ 0 & -1 \end{pmatrix}$

Game instructions and rules

You will need

- One copy of the transformation board.
- A pack of transformation cards.
- A counter.
- Paper and a pen to record scores.

How to play

1. Put the counter on the start triangle marked with an S.
2. Shuffle the cards face down and deal four to each player.
3. Place the rest of the pack face down on the table in an "unused pile".
4. Players take it in turns to move, for each turn they must try to move the counter from its current position to another triangle on the board using one of their transformation cards. (See variants.)
5. When a player has moved, they must throw away the used card onto a "discard" pile and collect a new card from the unused pile.
6. If a player cannot move they may discard one of their cards and collect a new one from the unused pile.
7. The cards are graded according to difficulty. For each move using a card A scores 4 points, a move using a card B scores 3 points, a move using a card C scores 2 points and a move using a card D scores 1 point.
8. If using transformation cards B, C or D players must describe in full the transformation they are doing, if they fail to do so they do not score the points.
9. If a player makes a mistake, the counter returns to its previous position and the player misses a turn and loses their card.
10. The game ends when the unused pile is empty and no more moves are possible.
11. The player with the highest score wins.

Variants

To make the game a little easier/faster to play deal five cards to each player.

You could make it harder by allowing combinations of transformations (step **4**); two or more cards per turn could be used. If you use more than one card, you could decide that the intermediate positions of the transforming triangle do not need to be marked on the game board.

Transformation game board

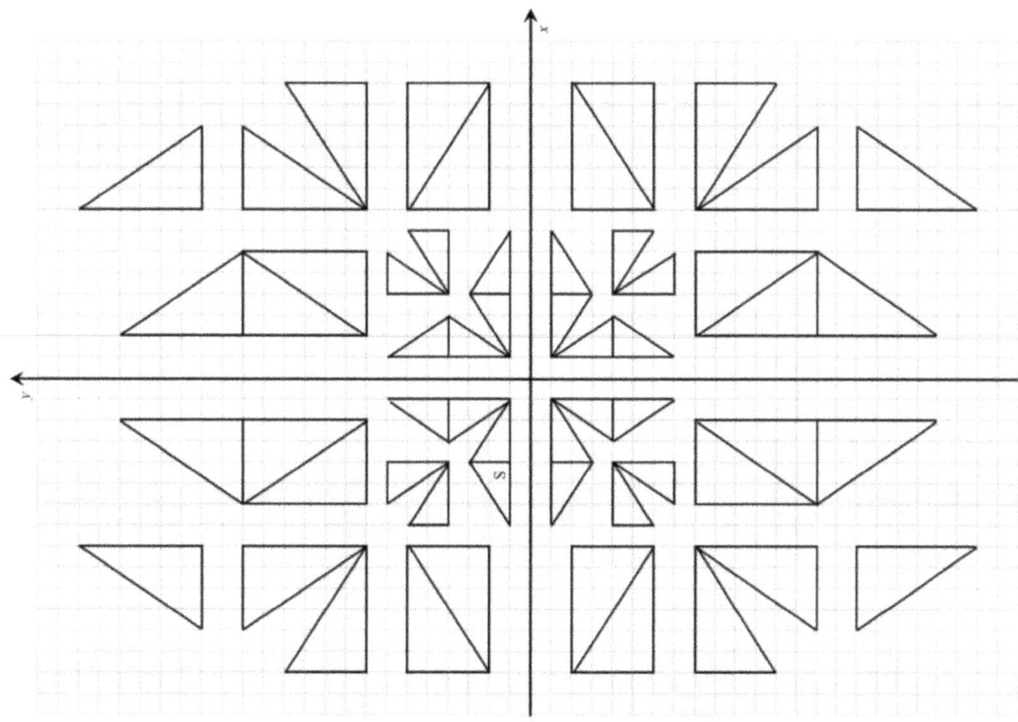

Game cards – set 1

Larger size game cards are available in the online resources.

CARD A
ROTATION
Centre: origin
Angle: 90°
Direction: clockwise

CARD A
ROTATION
Centre: origin
Angle: 90°
Direction: anticlockwise

CARD A
ROTATION
Centre: origin
Angle: 270°
Direction: clockwise

CARD A
ROTATION
Centre: origin
Angle: 270°
Direction: anticlockwise

CARD A
ROTATION
Centre: origin
Angle: 180°
Direction: not applicable

CARD B
ROTATION
Centre: origin
Angle: *player to say*
Direction: clockwise

CARD B
ROTATION
Centre: origin
Angle: *player to say*
Direction: *player to say*

CARD C
ROTATION
Player to describe in full

Game cards – set 2

Larger size game cards are available in the online resources.

CARD C
ROTATION
Player to describe in full

CARD C
ROTATION
Player to describe in full

CARD B
ROTATION
Centre: *player to say*
Angle: 90°
Direction: *player to say*

CARD B
ROTATION
Centre: *player to say*
Angle: *player to say*
Direction: clockwise

CARD C
REFLECTION
Player to describe in full

CARD C
REFLECTION
Player to describe in full

CARD C
REFLECTION
Player to describe in full

CARD A
REFLECTION
In the y-axis

Game cards – set 3

Larger size game cards are available in the online resources.

CARD A
REFLECTION
In the x-axis

CARD A
REFLECTION
In the y-axis

CARD A
REFLECTION
In the x-axis

CARD A
REFLECTION
In the line $x = 0$

CARD A
REFLECTION
In the line $y = 0$

CARD A
REFLECTION
In the line $y = x$

CARD A
REFLECTION
In the line $y = x$

CARD A
REFLECTION
In the line $y = -x$

Game cards – set 4

Larger size game cards are available in the online resources.

CARD A
REFLECTION
In the line $y = -x$

CARD B
REFLECTION
In the line $y =$ *player to say*

CARD B
REFLECTION
In the line $x =$ *player to say*

CARD A
TRANSLATION
Through the vector
$\begin{pmatrix} 3 \\ 3 \end{pmatrix}$ or $\begin{pmatrix} -3 \\ -3 \end{pmatrix}$

CARD A
TRANSLATION
Through the vector
$\begin{pmatrix} 6 \\ 0 \end{pmatrix}$ or $\begin{pmatrix} -6 \\ 0 \end{pmatrix}$

CARD B
TRANSLATION
Through the vector $\begin{pmatrix} a \\ b \end{pmatrix}$
where both a and b are positive

CARD B
TRANSLATION
Through the vector $\begin{pmatrix} a \\ b \end{pmatrix}$
where one of a and b is negative and one is positive

CARD C
TRANSLATION
Player to describe in full

Game cards – set 5

Larger size game cards are available in the online resources.

CARD C
TRANSLATION
Player to describe in full

CARD D
ANY TRANSFORMATION
Player to describe in full

CARD D
ANY TRANSFORMATION
Player to describe in full

CARD D
ANY TRANSFORMATION
Player to describe in full

CARD A
ENLARGEMENT
Centre: origin
Scale Factor: 2

CARD A
ENLARGEMENT
Centre: origin
Scale Factor: 2

CARD A
ENLARGEMENT
Centre: origin
Scale Factor: $\frac{1}{2}$

CARD A
ENLARGEMENT
Centre: origin
Scale Factor: $\frac{1}{2}$

Game cards – set 6

Larger size game cards are available in the online resources.

CARD A
ENLARGEMENT
Centre: origin
Scale Factor: –2

CARD A
ENLARGEMENT
Centre: origin
Scale Factor: –1

CARD B
ENLARGEMENT
Centre: origin
Scale Factor: *player to say*

CARD B
ENLARGEMENT
Centre: origin
Scale Factor: *player to say*

CARD C
ENLARGEMENT
Player to describe in full

CARD C
ENLARGEMENT
Player to describe in full

CARD C
ENLARGEMENT
Player to describe in full

CARD B
ENLARGEMENT
Centre: *player to say*
Scale Factor: 2

Surface area and volume crossword

Teacher notes

This activity is for students who already know most surface area and volume facts. It is a revision activity covering different areas of the syllabus including: volume and surface area of prisms, nets, capacity and units of volume and capacity. This activity focuses on some of the key vocabulary from the syllabus.

Introductory activity

There should be little need for any introduction, as this is intended as a revision exercise

Crossword activity

Print out one copy of the crossword grid and clues for your students, or pairs of students, ask them to complete the crossword by filling in one letter per box.

Differentiation

Ask students to complete the crossword against the clock. Alternatively ask them if they can devise alternative clues for some of the words.

Alternative approaches

Ask students to be creative to produce their own crossword using the blank crossword grid provided in the online resources.

Answers

			[1] M	I	L	L	I	L	I	T	R	E	S		
			I												
	[2] V	O	L	U	M	E			[3] C	I	R	[4] C	L	E	S
			L									R			
[5] R	A	D	I	U	[6] S				[7] C			O			
			O		U				A			S			[8] C
			N		R				[9] P	R	I	S	M		E
	[10] H				F				A			S			N
	E				A				C		[11] N	E	T		T
	I				[12] C	U	B	O	I	D		C			I
	G		[13] C		E				T			T			M
[14] T	H	O	U	S	A	N	D		Y			I			E
	T		R		R							O			T
			V		E			[15] C	Y	L	I	N	D	E	R
			E		A										E
			D												

Surface area and volume crossword

			1												
	2								3		4				
5					6				7						
															8
									9						
	10														
											11				
					12										
			13												
14															
								15							

Across

1 1 litre = 1000 ....

2 $\pi r^2 h$ is the formula for the ...... of a cylinder.

3 The net of a cylinder is made up of one rectangle and two .... , (we use it to help find the surface area).

5 The curved surface area of a cylinder can be found by calculating $2 \times \pi \times h \times$ ......

9 A three-dimensional shape with a uniform cross-section.

11 The diagram below is the .... of a cube.

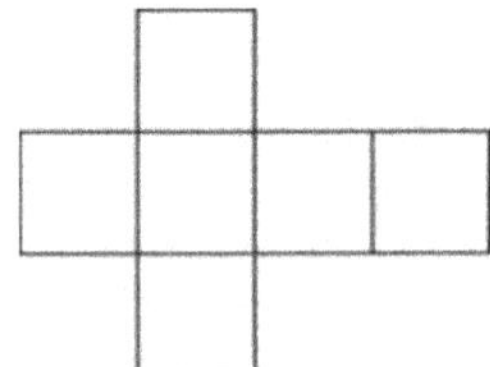

12 A ...... is a prism with a rectangular cross-section.

14 One cubic centimetre = one .... cubic millilitres.

15 A prism with a circular cross-section.

Down

1 One cubic metre = one .... cubic centimetres.

4 The volume of a prism is length × area of ....

6 The sum of the area of all the faces of a three-dimensional shape.

7 The amount that a container can hold; units for this are ml or l.

8 1 millilitre is the same as one cubic ....

10 Volume of a cuboid is length × width × ....

13 The surface of a cylinder consists of two circles and a rectangle. The rectangle is the .... surface of the cylinder.

Probability always true, never true or sometimes true card sort

Teacher notes

There are many misconceptions associated with probability. One of the most common errors is not considering all possible outcomes. For example, when tossing two coins students will often think there are three outcomes; two heads, two tails or one head and one tail, each with the probability of $\frac{1}{3}$, rather than the correct outcomes HH, HT, TH, TT, each with the probability of $\frac{1}{4}$. This activity is intended as a short starter task to check if students have any misconceptions, so you could complete this before starting probability work, or you could use it as a plenary at the end of the section in the student book.

Introductory activity

Little introduction should be necessary.

Card sorting activity

Print out the sort cards and cut them up, ask students to work in small groups and give each group a set of these cards. You could ask students simply to sort the cards into three piles: one pile of '**always true**', one pile of '**never true**' and one pile of '**sometimes true**' statements.

If you want to extend the activity ask the students to give the reason why they have placed the cards in each pile.

Differentiation

You can also differentiate by leaving some difficult cards out for less able students or by adding harder cards to extend the more able students. You can allow students a "**don't know**" pile. These can be sorted when you move around the class.

Alternative approaches

You could extend the task further, by asking students to write some examples of their own, using the blank cards from the online resources. The students can choose to do easy examples or really difficult ones. You could give them the incentive to find a difficult one by telling them their card will be given to another group to be classified.

Answers

Always True	
Statement	**Reason**
When you roll two fair six-sided dice it is harder to roll a total of 5 than a total of 7.	You can get a 5 by rolling 1, 4 or 4, 1 or 2, 3 or 3, 2 You can get a 7 by rolling 1, 6 or 2, 5 or 3, 4 or 4, 3 or 5, 2 or 6, 1 therefore a 7 is easier to roll.
In a class of 30 students the probability of two students being born in the same month is 1.	There are only 12 months in a year so it is certain that at least 2 students will be born in the same month.
Never True	
Statement	**Reason**
A bag contains cubes numbered 1 to 20. Four cubes are picked at random. The four cubes numbered 3, 7, 14, 19 are more likely to get picked than the cubes numbered 1, 2, 3 and 4.	Any combination of four numbers is equally likely.

The probability of there being two girls in a family with four children is $\frac{1}{2}$.	Possible outcomes for four children are: BBBB, BBBG, BBGB, **BBGG**, BGBB,**BGBG**, **BGGB**, BGGG, GBBB, **GBBG**, **GBGB**, GBGG, **GGBB**, GGBG, GGGB, GGGG Those in **bold** have exactly two girls, so the probability $= \frac{6}{16} = \frac{3}{8}$. If the question had asked for **at least** 2 girls that would be the underlined ones which is $\frac{11}{16}$, neither of these are $\frac{1}{2}$.
In a family with two children there are three possible outcomes: Two boys, Two girls, One boy and one girl. Therefore the probability of two boys is $\frac{1}{3}$	In a family with two children there are four possible outcomes: BB, BG, GB, GG as the order in which the children are born needs to be taken into account. So the probability of two boys is $\frac{1}{4}$.
A fair coin is tossed with two possible outcomes, heads or tails. The coin is tossed 4 times. If the first three tosses are heads then the last toss is more likely to be a tail.	It doesn't matter what the first three outcomes are, you are still equally likely to get a head as you are a tail on the fourth toss.
When you roll a fair six-sided die it is harder to roll a six than a two.	All scores are equally likely, all have a probability of $\frac{1}{6}$.
When you roll two fair six-sided dice it is harder to roll a total of 7 than a double.	You can get a 7 by rolling 1, 6 or 2, 5 or 3, 4 or 4, 3 or 5, 2 or 6, 1 to get a double you need 1, 1 or 2, 2 or 3, 3 or 4, 4 or 5, 5 or 6, 6. Both events have six outcomes so are equally likely.
Sometimes true	
Statement	**Reason**
In a true or false quiz with 20 questions you will get 10 correct if you answer "true" for every question.	This could be true however you can't tell. It will only be true if 10 questions have the correct answer as true and 10 have the correct answer as false.
In a true or false quiz with 20 questions you will get 10 correct if you guess the answer to every question.	This is unlikely but possible. If you had no idea about any question then you do have an even chance of getting each question correct therefore 10 marks. However in reality you may well know the answer to some of the questions.
In a football match there are three possible outcomes: Win, Lose, Draw. Therefore the probability of losing is $\frac{1}{3}$.	The probabilities of each outcome depend on the relative abilities of the two teams, match conditions, etc. and are not usually equally likely.
In a class of 30 students the probability of two students being born in January is 1.	We know at least 2 students must be born in the same month, but that month doesn't have to be January. For example 29 students could be born in December and 1 in January. Unlikely but possible.

Probability always true, never true or sometime true sort cards

A bag contains cubes numbered 1 to 20. Four cubes are picked at random. The four cubes numbered 3, 7, 14, 19 are more likely to get picked than the cubes numbered 1, 2, 3 and 4.	When you roll two fair six-sided dice it is harder to roll a total of 5 than a total of 7.
The probability of there being two girls in a family with four children is $\frac{1}{2}$.	In a family with two children there are three possible outcomes: • Two boys • Two girls • One boy and one girl Therefore the probability of two boys is $\frac{1}{3}$.
A fair coin is tossed with two possible outcomes, heads or tails. The coin is tossed 4 times. If the first three tosses are heads then the last toss is more likely to be a tail.	When you roll a fair six-sided die it is harder to roll a six than a two.
In a true or false quiz with 20 questions you will get 10 correct if you answer "true" for every question.	In a true or false quiz with 20 questions you will get 10 correct if you guess the answer to every question.
In a class of 30 students the probability of two students being born in the same month is 1.	In a football match there are three possible outcomes: • Win • Lose • Draw Therefore the probability of losing is $\frac{1}{3}$.
In a class of 30 students the probability of two students being born in January is 1.	When you roll two fair six-sided dice it is harder to roll a total of 7 than a double.

Percentage increase and decrease card sort

Teacher notes

The purpose of this card grouping activity is to find the links between percentages, decimals and fractions, and to be able to represent percentage increase and decrease as multiplication.

There are many different approaches to percentage questions. Students can use this card grouping activity to look for the links between the different approaches. The cards also include the descriptions of the fractions in words as well as figures. This will be helpful for students who do not have English as their first language. Many students wrongly believe that an increase of 50% followed by a decrease of 50% takes them back to the original value and they are confused by the inverse relationship between percentage increase and percentage decrease. One of the aims of this activity is to eliminate such misconceptions.

Introductory activity

Discuss with students the equivalences between percentages, decimals and fractions. For example, to **increase by 10%** you have the original 100% plus 10% more which is 110% and this is where the **× 1.1** comes from – the decimal equivalent to 110%. Students can also use a fractional approach as 10% is equivalent to **one tenth**, and going up by one tenth is the original 1 plus $\frac{1}{10}$ which explains the $\times\frac{11}{10}$.

Ask students to increase \$100 by 50%. They should get \$150. Then ask them to reduce \$150 by 50% and they should get \$75. Draw the following diagram on the board and ask them for the missing percentage.

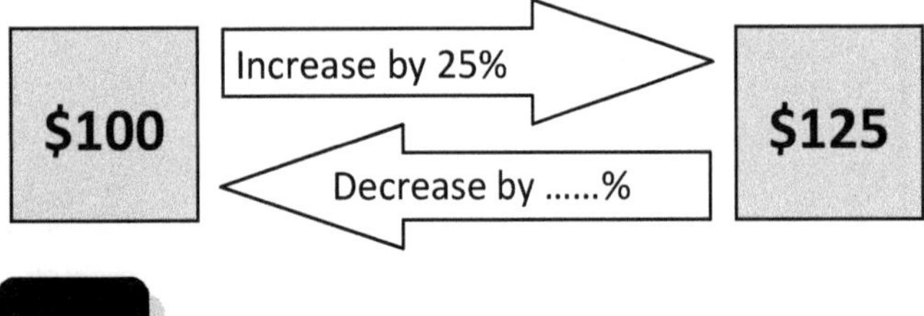

If students need help with this, point out to them that $\frac{1}{5}$ of \$125 is \$25, so a decrease of $\frac{1}{5}$ is equivalent to a decrease of 20%, which will take them back to \$100.

Card sorting activity

Print out the sort cards and cut them up before handing them out. Explain to students that you have a series of cards that are to be sorted into groups of four, give them the example below to demonstrate a complete set. Explain that there are a mix of increase and decrease sets and that there are ten sets in total.

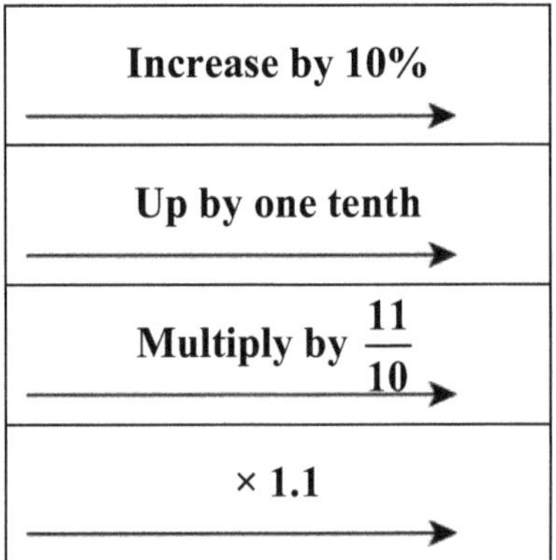

Differentiation

To make the task harder, once students have grouped the cards into sets of four you could ask them to pair the sets up. Students will need to start with a percentage increase and then decide what percentage decrease set will reverse the increase and take them back to the original. If they struggle with this then the following example may help.

Example	**Reversed**
Increase by 10%,	Decrease by $9\frac{1}{11}$%,
Up by one tenth,	Down by one eleventh,
$\times\frac{11}{10}$, $\times$ 1.1	$\times\frac{10}{11}$, $\times$ $0.\dot{9}\dot{0}$

Hopefully students will notice that the easiest way to pair the cards up is to find fractions that are reciprocals of each other.

Alternative approaches

Allow students more freedom: simply ask them to sort the cards, without explaining how many there are in each set. You could include a creative approach by asking students to add a further set of their own and the reverse set.

You could also print out some money cards, available in the online resources, make up some of your own, or get the students to make up some, and ask students to sort the cards as in the following example.

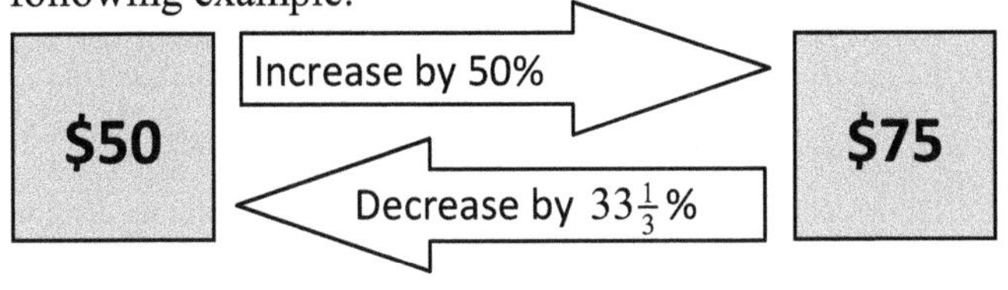

Answers

The sets are arranged in 'reverse' pairs, 1 and 6, 2 and 7, etc.

Set 1	**Set 6**
Increase by 25%,	Decrease by 20%,
Up by one quarter,	Down by one fifth,
$\times\frac{5}{4}$, $\times 1.25$	$\times\frac{4}{5}$, $\times 0.8$
Set 2	**Set 7**
Increase by 80%,	Decrease by $44\frac{4}{9}$%,
Up by four fifths,	Down by four ninths,
$\times\frac{9}{5}$, $\times 1.8$	$\times\frac{5}{9}$, $\times 0.\dot{5}$
Set 3	**Set 8**
Increase by 50%,	Decrease by $33\frac{1}{3}$%,
Up by one half,	Down by one third,
$\times\frac{3}{2}$, $\times 1.5$	$\times\frac{2}{3}$, $\times 0.\dot{6}$
Set 4	**Set 9**
Increase by $66\frac{2}{3}$%,	Decrease by 40%,
Up by two thirds,	Down by two fifths,
$\times\frac{5}{3}$, $\times 1.\dot{6}$	$\times\frac{3}{5}$, $\times 0.6$
Set 5	**Set 10**
Increase by 100%,	Decrease by 50%,
Double,	Down by one half,
$\times\frac{2}{1}$, $\times 2$	$\times\frac{1}{2}$, $\times 0.5$

Possible answers to the alternative approach are:

$100	→ increase by 25% ← decrease by 20%	$125
$100	→ increase by 80% ← decrease by $44\frac{4}{9}$%	$180
$100	→ increase by 100% ← decrease by 50%	$200
$36	→ increase by $66\frac{2}{3}$% ← decrease by 40%	$60
$48	→ increase by 25% ← decrease by 20%	$60
$35	→ increase by 80% ← decrease by $44\frac{4}{9}$%	$63
$300	→ increase by $66\frac{2}{3}$% ← decrease by 40%	$500
$180	→ increase by 50% ← decrease by $33\frac{1}{3}$%	$270

Percentage increase and decrease sort cards

Increase by 25% →	← Decrease by 20%
Up by one quarter →	← Down by one fifth
Multiply by $\frac{5}{4}$ →	← Multiply by $\frac{4}{5}$
× 1.25 →	← × 0.8
Increase by 80% →	← Decrease by $44\frac{4}{9}$%
Up by four fifths →	← Down by four ninths
Multiply by $\frac{9}{5}$ →	← Multiply by $\frac{5}{9}$
× 1.8 →	← × $0.\dot{5}$
Increase by 50% →	← Decrease by $33\frac{1}{3}$%
Up by one half →	← Down by one third

Multiply by $\frac{3}{2}$ →	← Multiply by $\frac{2}{3}$
× 1.5 →	← × $0.\dot{6}$
Increase by $66\frac{2}{3}$% →	← Decrease by 40%
Up by two thirds →	← Down by two fifths
Multiply by $\frac{5}{3}$ →	← Multiply by $\frac{3}{5}$
× $1.\dot{6}$ →	← × 0.6
Increase by 100% →	← Decrease by 50%
Double →	← Down by one half
Multiply by $\frac{2}{1}$ →	← Multiply by $\frac{1}{2}$
× 2 →	← × 0.5

Larger size jigsaw pieces are available in the online resources.

Bearings drawing

Teacher notes

This activity can be used as a starter, a plenary, a homework task or a revision exercise. This fun task revises the basics of bearings work, in particular that the angle should always be three figures and measured clockwise from the North line.

Introductory activity

Although this is quite a basic exercise, you may wish to revise the following:

- An isometric grid is made from equilateral triangles with interior angles of 60°.
- A bearing is always measured in a clockwise direction from the North line and is always three figures.

Example 1

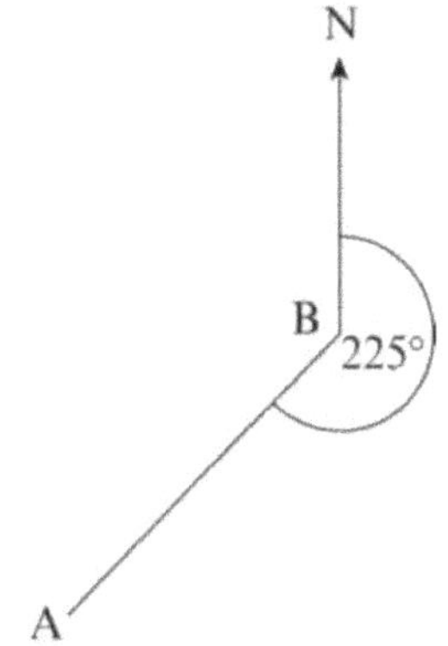

The bearing of A from B is 225°

Example 2

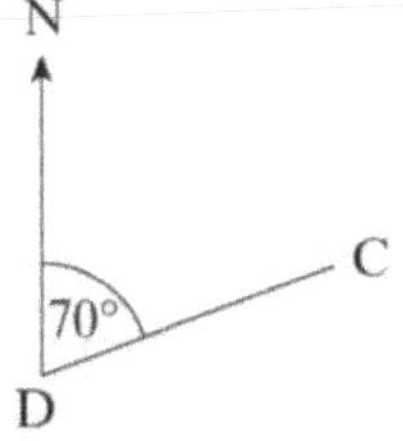

The bearing of C from D is 070°

Bearing picture

Give each student a copy of the worksheet. Ask them to start at point number 1 on the isometric grid, and to draw a trail of lines, with lengths and bearings as given in the corresponding table, each of which follows on from the end of the previous line. In doing this, students will draw each part of the picture. 1 unit of length is the distance from one dot to another on the isometric grid. For each numbered North line there is a new table to start a new drawing.

Differentiation

To make the task easier you can tell students that the image should be a horse, or to make it slightly more difficult you can leave them to discover this for themselves. You could also tell weaker students that the image is symmetrical. To make the task harder, ask the students to add extra features to the drawing (e.g. a pattern on the cheeks of the horse) along with the instructions for drawing any additional lines. They could ask their classmate to follow their instructions to draw the extra features.

Alternative approaches

Ask students to create their own drawing on an isometric grid along with the instructions for drawing all of the lines. Students could also create their own drawing, and instructions, on plain paper so they can use any angle and any length line.

Bearings drawing

Do a different drawing for each number. Start at the North line and join the dots according to the bearings and lengths given. Start each new line where the previous line ends.

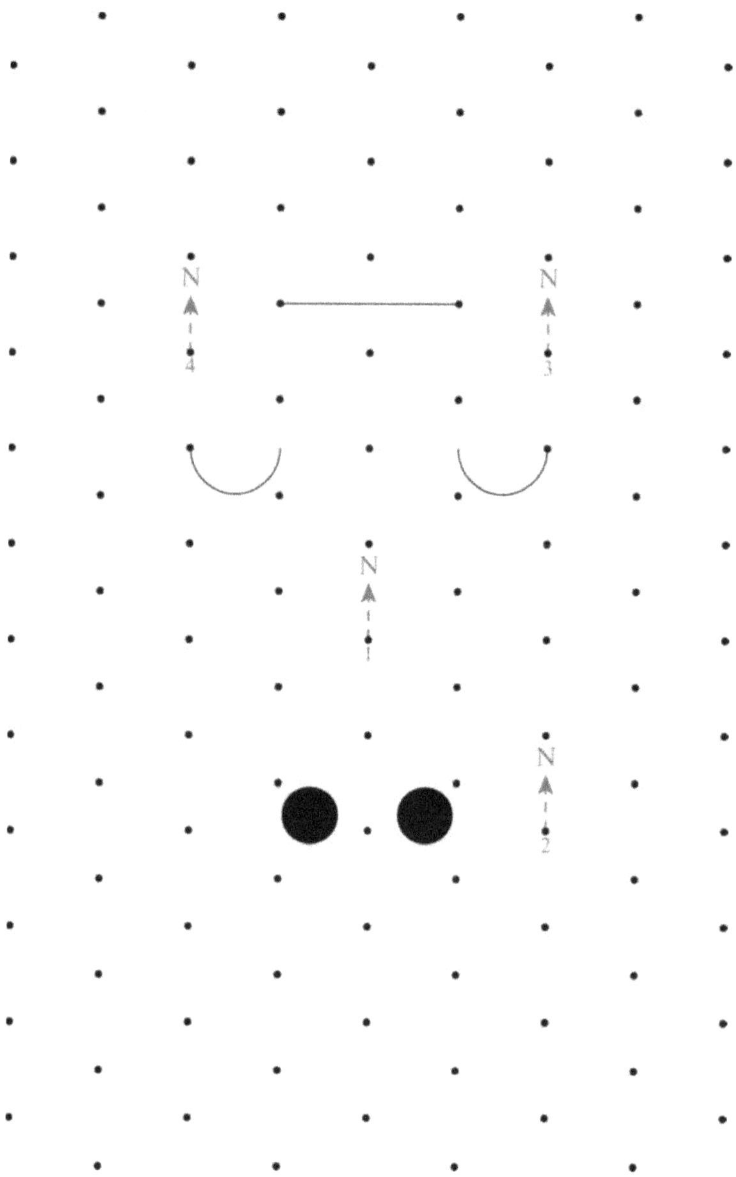

1.

Bearing	Length
240°	2
180°	2
120°	2
060°	2
000°	2
300°	2
000°	3
300°	1
240°	2
180°	4
120°	1

2.

Bearing	Length
060°	1
000°	4
300°	2
240°	1

3.

Bearing	Length
060°	1
000°	2
240°	1
180°	2

4.

Bearing	Length
000°	2
300°	1
180°	2
120°	1

Answer

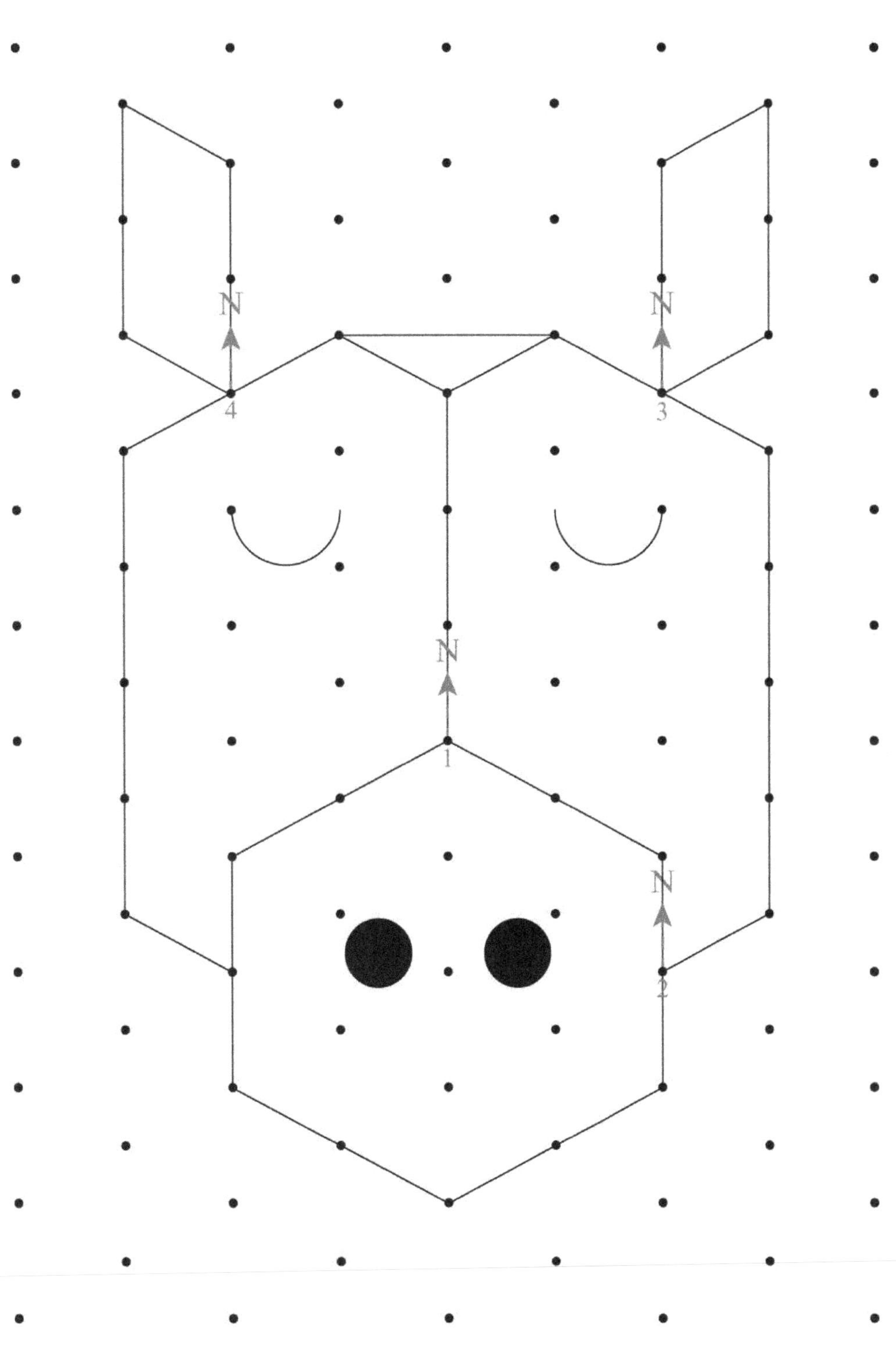

Trigonometry crossnumber

Teacher notes

This activity is for students who already know the basics of the sine, cosine and tangent ratios. It can be used as a fun revision session or a quick homework activity. Students will need to know how to calculate missing side lengths and missing angles in basic right-angled triangles only. There are no complicated questions as this is intended only as revision of the basics.

Introductory activity

There should be little need for any introduction, as this activity is intended to follow directly from basic trigonometry; before starting more complex work such as 3D trigonometry, angles of elevation and depression and trigonometry questions in context. You may want work through these examples.

Labelling the triangle and the three ratios:

$$\text{Sin } \theta = \frac{\text{Opp}}{\text{Hyp}}$$

$$\text{Cos } \theta = \frac{\text{Adj}}{\text{Hyp}}$$

$$\text{Tan } \theta = \frac{\text{Opp}}{\text{Adj}}$$

(SOH CAH TOA)

Example 1

Finding a missing angle: find x to 1 d.p.

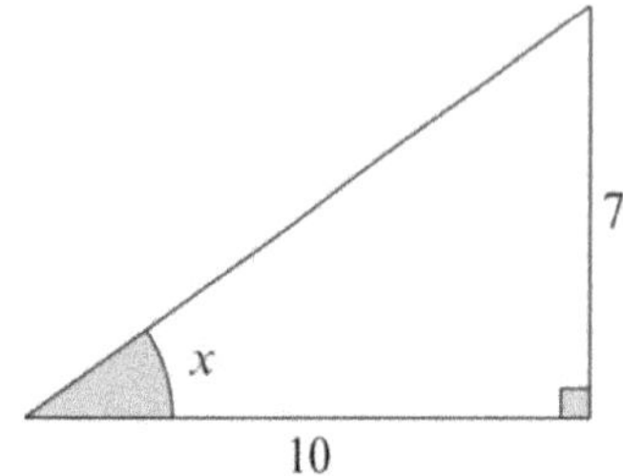

$$\tan \theta = \frac{\text{opp}}{\text{adj}} \quad \text{so} \quad \tan x = \frac{7}{10} = 0.7$$

$x = \tan^{-1}(0.7) = 34.9920202\ldots$

$= 35.0° \quad$ (to 1 d.p.)

Example 2

Finding a missing side length: find x to 3 s.f.

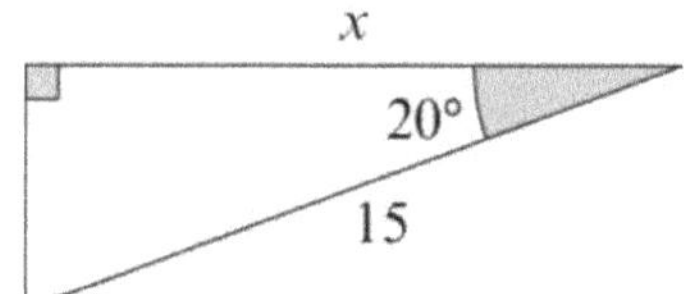

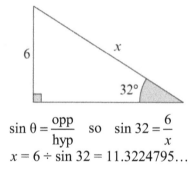

$$\cos \theta = \frac{\text{adj}}{\text{hyp}} \quad \text{so} \quad \cos 20 = \frac{x}{15}$$

$x = 15 \times \cos 20 = 14.0953893\ldots$

$= 14.1 \quad$ (to 3 s.f.)

Example 3

Finding a missing side length: find x to 3 s.f.

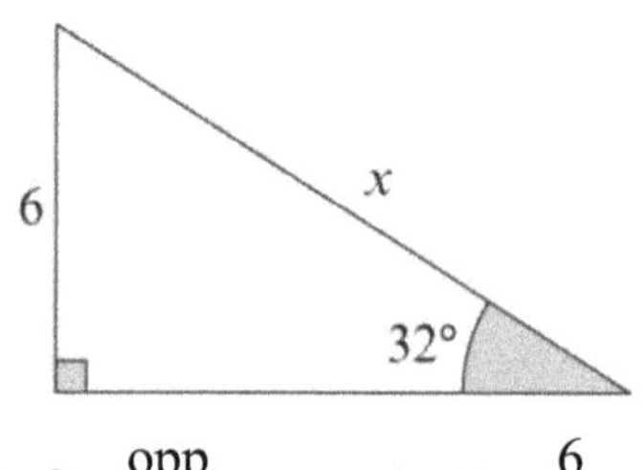

$$\sin \theta = \frac{\text{opp}}{\text{hyp}} \quad \text{so} \quad \sin 32 = \frac{6}{x}$$

$x = 6 \div \sin 32 = 11.3224795\ldots$

$= 11.3 \quad$ (to 3 s.f.)

Exam questions usually ask for angles rounded to 1 d.p. and side lengths to 3 s.f. You may want to stress to your students that it is a good idea to write unrounded answers in their working before rounding.

Crossnumber activity

Print out one copy of the crossnumber grid and clues for your students or pairs of students. Ask them to complete the crossnumber by filling in one digit or **decimal point** per box.

Differentiation

Ask students to race to complete the crossnumber, or use the blank grid, available from the online resources, to make harder/easier clues of your own.

Alternative approaches

You could give the students a blank copy of the crossword, from the online resources, asking them to write the clues and the solutions, this is a challenging task.

Answers

			[1] 3				[2] 3	[3] 3	·	7			
			5					3					
			·					·					
[4] 3	[5] 2	·	7				[6] 2	1	·	[7] 8			
	8						2			·		[8] 1	
	·				[9] 1	5	·	7		7		3	
[10] 1	2	·	[11] 4		5		4			[12] 7	0	·	5
4			·		·							7	
·		[13] 4	5	·	6								
9			9										

Across

2 $\tan x = 8 \div 12$

$x = \tan^{-1}(8 \div 12) = 33.7$ (to 1.d.p.)

4 $\sin x = 10 \div 18.5$

$x = \sin^{-1}(10 \div 18.5) = 32.7$ (to 1.d.p.)

6 $\sin x = 4 \div 10$

$x = \tan^{-1}(4 \div 10) = 21.8$ (to 1.d.p.)

9 $\cos 40 = 12 \div x$

$x = 12 \div \cos 40 = 15.7$ (to 3.s.f.)

10 $\tan 22 = 5 \div x$

$x = 5 \div \tan 22 = 12.4$ (to 3.s.f.)

12 $\cos x = 6 \div 8$

$x = \cos^{-1}(6 \div 18) = 70.5$ (to 1.d.p.)

13 $\cos x = 14 \div 20$

$x = \cos^{-1}(14 \div 20) = 45.6$ (to 1.d.p.)

Down

1 $\sin x = 7 \div 12$

$x = \sin^{-1}(7 \div 12) = 35.7$ (to 1 d.p.)

3 $\tan x = 2.8 \div 4.3$

$x = \tan^{-1}(2.8 \div 4.3) = 33.1$ (to 1 d.p.)

5 $\tan 28 = 15 \div x$

$x = 15 \div \tan 28 = 28.2$ (to 3 s.f.)

6 $\cos 17 = x \div 23.4$

$x = 23.4 \times \cos 17 = 22.4$ (to 3 s.f.)

7 $\sin 20 = 3 \div x$

$x = 3 \div \sin 20 = 8.77$ (to 3 s.f.)

8 $\sin 41 = 9 \div x$

$x = 9 \div \sin 41 = 13.7$ (to 3 s.f.)

9 $\cos 30 = x \div 18$

$x = 18 \times \cos 30 = 15.6$ (to 3 s.f.)

10 $\tan 43 = x \div 16$

$x = 16 \times \tan 43 = 14.9$ (to 3 s.f.)

11 $\sin 35 = x \div 8$

$x = 8 \times \sin 35 = 4.59$ (to 3 s.f.)

Trigonometry crossnumber

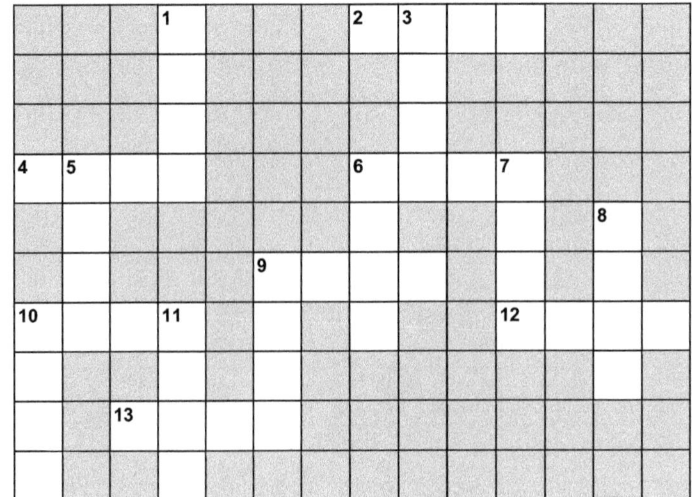

Across

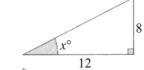

2 8, 12, $x°$ Find x to 1 d.p.

4 10, 18.5, $x°$ Find x to 1 d.p.

6

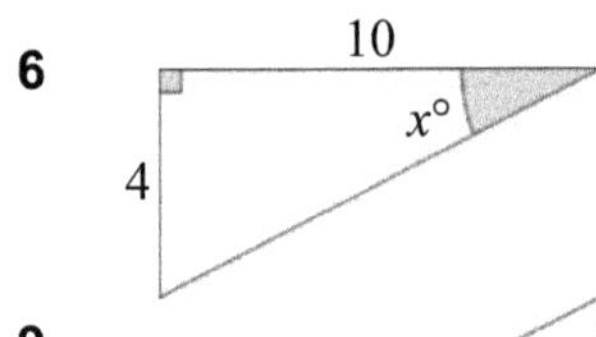

Find x to 1 d.p.

9

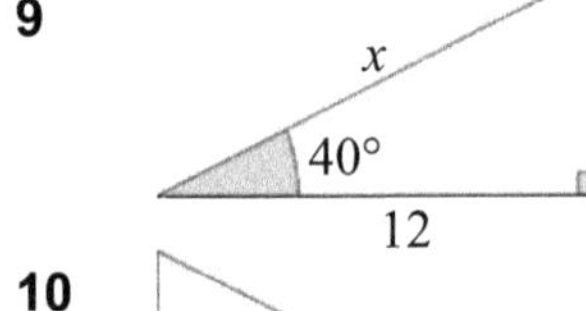

Find x to 3 s.f.

10 5, 22°, x Find x to 3 s.f.

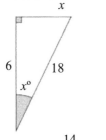

12 6, 18, $x°$ Find x to 1 d.p.

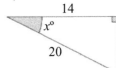

13 14, $x°$, 20 Find x to 1 d.p.

Down

1

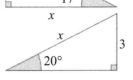

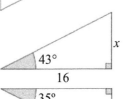

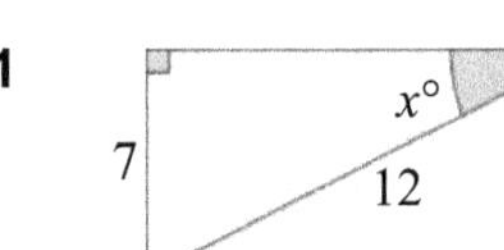

Find x to 1 d.p.

3 $x°$, 2.8, 4.3 Find x to 1 d.p.

5 28°, 15, x Find x to 3 s.f.

6 23.4, 17°, x Find x to 3 s.f.

7 x, 20°, 3 Find x to 3 s.f.

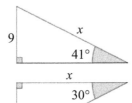

8 9, x, 41° Find x to 3 s.f.

9 x, 30°, 18 Find x to 3 s.f.

10 43°, x, 16 Find x to 3 s.f.

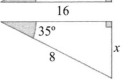

11 35°, 8, x Find x to 3 s.f.

Teacher notes

This activity is a revision activity on scatter diagrams and correlation. For a collective memory activity students are shown a poster for a short period of time, and they must try to reproduce as much of the poster as they can. There is an opportunity to highlight common misconceptions, encourage teamwork and make maths lessons, and in particular note taking or revising, more interesting and fun.

Introductory activity

There should be little need for any introduction, as this is intended as a revision exercise.

Collective memory activity

You will need to have on your desk a copy of the collective memory poster (preferably double the size, A3, using the colour version from the online resources). You will need a stopswatch, (you can find one online at http://www.online-stopwatch.com/). Divide the class into groups of four students. Give each group a blank sheet of A3 paper and coloured pens/pencils to match those used in the memory poster. Tell students that during this activity you expect them to communicate with their teammates and devise strategies for how to approach this task.

Give each student in the group a number from 1 to 4). To start, all students numbered 1 come to the front of the class and have 30 seconds to view the poster. Then they return to their group and have one minute to produce as much of the poster as they can remember, as well as communicate with their teammates what areas the next person should concentrate on. Then all students numbered 2 view the poster for 30 seconds and have one minute back with their team. This continues until each team member has seen the poster twice. Allow two minutes at the end for the group to finish off their poster.

At the end of the activity reveal the poster to the class and discuss the activity considering some of the following points.

- Which parts of the poster were easier to recreate, which were more difficult and why?
- What are the most important features of the poster?
- How well did they work as a team?

Differentiation

To make this task harder do not allow the student who viewed the poster to draw anything, instead the next numbered person in the team has to do the drawing. This means even more cooperation and communication is required.

You could make it easier or harder by:

- adjusting the time for which students are allowed to view or draw
- allowing more or fewer than two views per student
- putting more or less information in the poster

Alternative approaches

You could display an electronic copy of the memory poster on the board to avoid the need for students to move around the classroom and the entire group can view the poster at the same time, which can speed up the activity. You can still keep the rule that only one student can write at a time and that they must take it in turns to do so.

You could ask students to work individually instead of in groups.

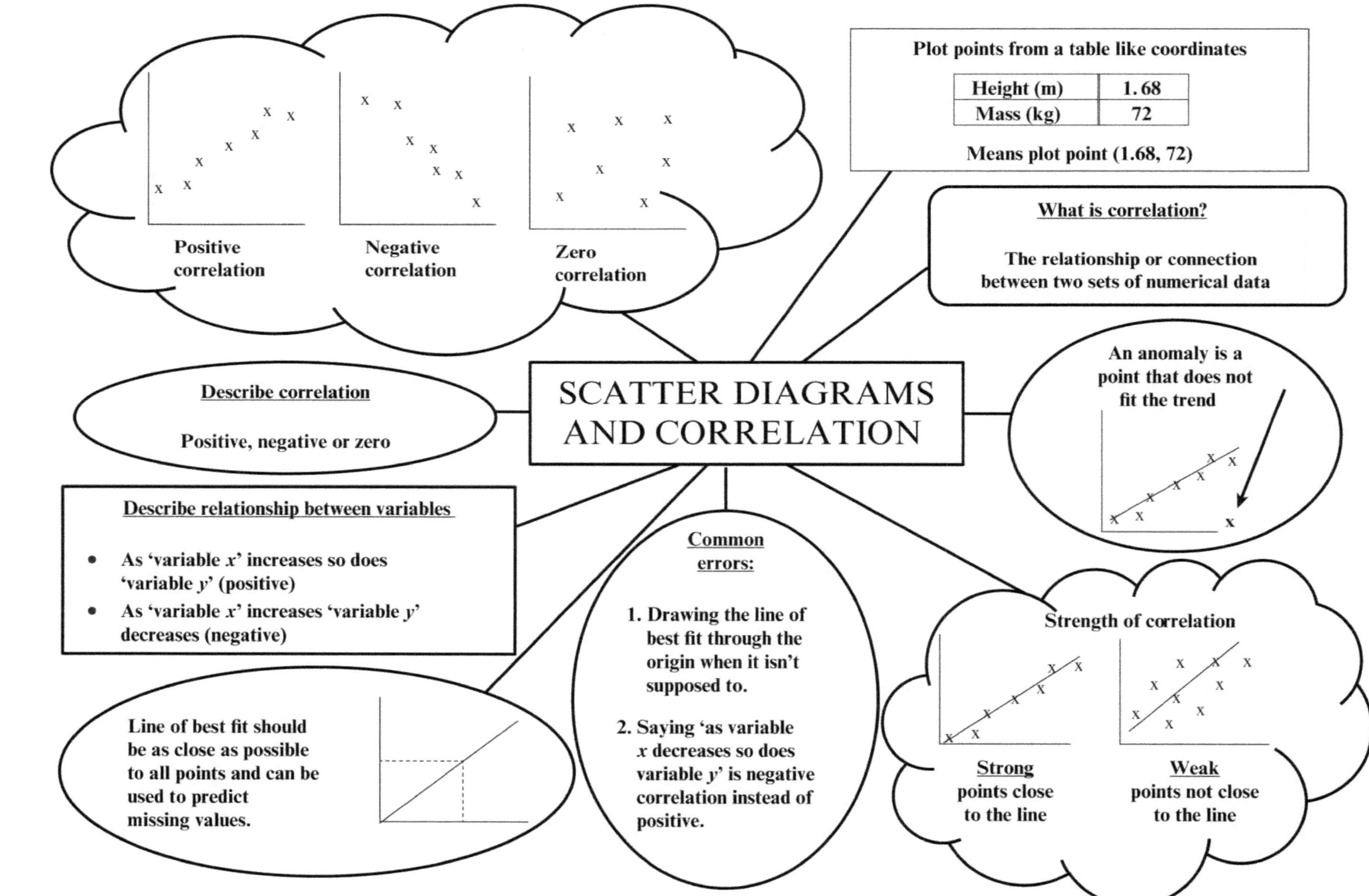
SCATTER DIAGRAMS AND CORRELATION
Positive correlation
Negative correlation
Zero correlation
Plot points from a table like coordinates
Height (m) | 1.68
Mass (kg) | 72
Means plot point (1.68, 72)
What is correlation?
The relationship or connection between two sets of numerical data
An anomaly is a point that does not fit the trend
Describe correlation
Positive, negative or zero
Describe relationship between variables
• As 'variable x' increases so does 'variable y' (positive)
• As 'variable x' increases 'variable y' decreases (negative)
Common errors:
1. Drawing the line of best fit through the origin when it isn't supposed to.
2. Saying 'as variable x decreases so does variable y' is negative correlation instead of positive.
Strength of correlation
Strong points close to the line
Weak points not close to the line
Line of best fit should be as close as possible to all points and can be used to predict missing values.

Direct and inverse proportion hidden treasure hunt

Teacher notes

This activity can be used as a starter, a plenary, or a revision exercise later on in the year. Students will need to understand the difference between direct and inverse proportion and be able to do associated questions, including practical, real-life examples and currency conversions.

The biggest problem in currency conversion work is that students divide when they are supposed to multiply or vice versa. This is either due to not reading the question carefully or a lack of understanding.

Introductory activity

If you are doing this as a plenary activity at the end of the chapter then little introduction will be necessary. However you may want to use this activity as an alternative to a student book exercise, in which case you may need to remind students of the following.

Direct proportion

As one value increases so does the other at the same rate. For example, if one dollar is worth 84 Japanese Yen then if you double the amount of dollars you double the amount of Japanese Yen it is worth, so \$2 = 168 Japanese Yen.

Inverse proportion

As one value increases the other value decreases at the same rate. For example a bag of food will feed 20 animals for 7 days, if you halve the amount of animals you double the amount of days the food will last, so if you only have 10 animals the same bag of food will last 14 days.

Hidden treasure hunt game

On the board draw a set of co-ordinate axes with both the x- and y-axes from 0 to 3.

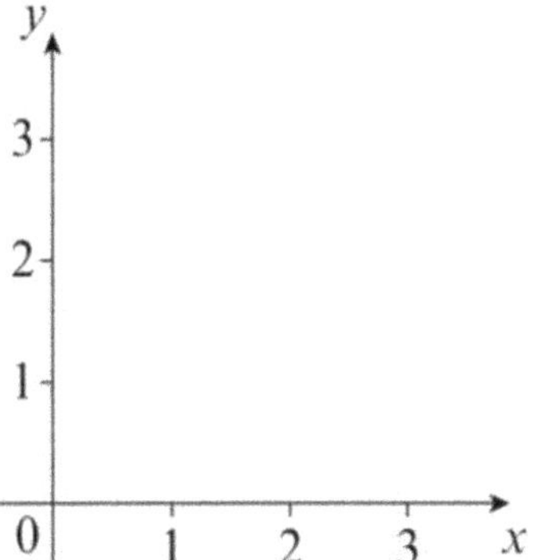

Have a pre-prepared list of questions on direct and inverse proportion. A suggested set of questions together with answers is supplied at the end.

Explain that this is a treasure map and at one of the co-ordinates there is buried treasure. Their aim is to try to find the treasure.

Invite students in turn to state co-ordinates and then answer the associated question.

If they get the answer right, tell the class if the treasure is buried there or not and if not mark off on the axes with a cross where their guess was.

If they get the answer wrong, do not tell them if the treasure is buried there or not and the class loses a life. The class has five lives.

Differentiation

You could reduce the number of lives to make this harder or increase the number of lives to make it easier. You may choose not to mark off where the students have guessed then they will have to pay close attention to where other students have already guessed or they will end up answering the same questions again.

Alternative approaches

Choose your own questions and write them in a copy of the blank grid from the online resources.

Questions and answers

Key: **D** = direct proportion **I** = inverse proportion

(0, 0) **D**	The cost of five apples is $3.95. What is the cost of three apples?	1 apple is $0.79, so 3 apples are **$2.37**	(2, 0) **D**	The exchange rate from dollars to GB£ was £1 = $1.32. At this rate how many pounds (£) would you get for $237.60?	237.60 ÷ 1.32 = **£180**
(0, 1) **I**	Eight men build a house in 6 weeks. How long will it take six men to build the same house?	Total weeks 8 × 6 = 48, so six men would take 48 ÷ 6 = **8 hours**	(2, 1) **D**	A car uses 18 litres of petrol to travel 135 km. If the car uses 10 litres of petrol how far will it travel?	1 litre gives 7.5 km, so 10 litres gives **75 km**
(0, 2) **I**	Toys are shared between 4 children so that they each have 6 toys. How many toys would each child have if there were 12 children?	Total toys 4 × 6 = 24, so 12 children would each get 24 ÷ 12 = **2 toys**	(2, 2) **D**	The exchange rate from dollars to Japanese Yen was $1 = 82.6 Yen. At this rate how many dollars would you get for 2065 Japanese Yen?	2065 ÷ 82.6 = **$25**
(0, 3) **I**	A bag of food will feed 15 animals for 6 days. How many days would the same bag of food last 18 animals?	Total days 15 × 6 = 90, so for 18 animals the food would last 90 ÷ 18 = **5 days**	(2, 3) **D**	Fernando earns $128.45 for 7 hours work. How much will he be paid for 3 hours work?	1 hour is worth $18.35 so 3 hours is worth **$55.05**
(1, 0) **D**	Sam earns $43.95 for 3 hours work. How much he will be paid for 10 hours work?	1 hour is worth $14.65 so 10 hours is worth **$146.50**	(3, 0) **D**	The cost of 7 pens is $8.47. What is the cost of four pens?	1 pen is $1.21, so 4 pens are **$4.84**
(1, 1) **D**	The exchange rate from dollars to GB£ was £1 = $1.27. At this rate how many dollars would you get for £70?	70 × 1.27 **= $88.90**	(3, 1) **I**	Four people take 6 hours to paint a house. How long will it take three people?	Total hours 4 × 6 = 24, so three people would take 24 ÷ 3 = **8 hours**
(1, 2) **I**	It takes 8 cleaners 10 hours to clean all the rooms in an office block. How long would it take 5 cleaners?	Total hours 8 × 10 = 80, so 5 cleaners would take 80 ÷ 5 = **16 hours**	(3, 2) **I**	Three men build a wall in 15 hours. How long will it take five men to build the same wall?	Total hours 3 × 15 = 45, so five men would take 45 ÷ 5 = **9 hours**
(1, 3) **D**	A car uses 15 litres of petrol to travel 120 km. If the car uses 20 litres of petrol how far will it travel?	1 litre gives 8 km, so 20 litres gives **160 km** (**or** 5 litres gives 40 km, so 20 litres gives 160 km)	(3, 3) **I**	Three sacks of potatoes will feed 15 people for 4 days. For how many days will three sacks of potatoes feed 12 people?	Total days 4 × 15 = 60, so it would last 12 people 60 ÷ 12 = **5 days**

Inverse functions jigsaw

Teacher notes

This jigsaw activity is an interesting group work alternative to a written exercise. The advantages of a jigsaw activity over a written exercise are that it creates opportunities for students to discuss their ideas, and they can find an alternative starting point if they are not sure how to do a particular question. Students are able to start with the answer and work backwards, which is sometimes an interesting way to tackle a question.

This activity requires students to be able to work out the inverse of given functions. Encourage students to do this by thinking of the function in the form y = something instead of $f(x)$ = something, then interchanging the x and y, rearranging to make y the subject and finally re-writing the y as $f^{-1}(x)$.

Introductory activity

You should explain to students that they will be able to do many of the questions in their heads, though for some they may need to jot down workings. Encourage students to discuss the order of operations so that they will be able to rearrange correctly. If you want to use this as a revision activity you may find that little more introduction is necessary. Otherwise the following example should be sufficient.

Find the inverse of the function $f(x) = 3x^2 - 5$

Step 1 $f(x) = 3x^2 - 5$ replace $f(x)$ with y

Step 2 $y = 3x^2 - 5$ interchange the x and y

Step 3 $x = 3y^2 - 5$ rearrange to make y the subject

Step 4 $x + 5 = 3y^2$

$$\frac{x+5}{3} = y^2$$

$$\sqrt{\frac{x+5}{3}} = y$$ replace the y with $f^{-1}(x)$

Step 5 $f^{-1}(x) = \sqrt{\frac{x+5}{3}}$

Jigsaw activity

Ask your students to work in small groups. Copy the jigsaw pieces onto card to make a complete set for each group. There are several different-shaped jigsaws in the Teacher Resource Kit so you should not tell the students what shape to expect. This jigsaw has blank edges so it is easier for your students to work out.

Differentiation

To make it easier you could simply ask students to do the jigsaw. To make it harder you could remove one or two of the pieces from the middle giving them blank pieces, and asking them to complete these pieces to fill in the gaps. Blank pieces can be found in the online resources.

Alternative approaches

You could copy blank jigsaw pieces from the online resources and ask students to use these pieces to make an entirely new jigsaw of their own, or ask them to consider extending this jigsaw. Note: if you intend asking them to extend this jigsaw be prepared for them to write on the printed jigsaw as it has blank edges. You could ask the students some additional questions such as.

1 Make a note of functions that are their own inverses.

2 Identify which functions are easier to work out the inverse functions and what makes them easier.

Answers

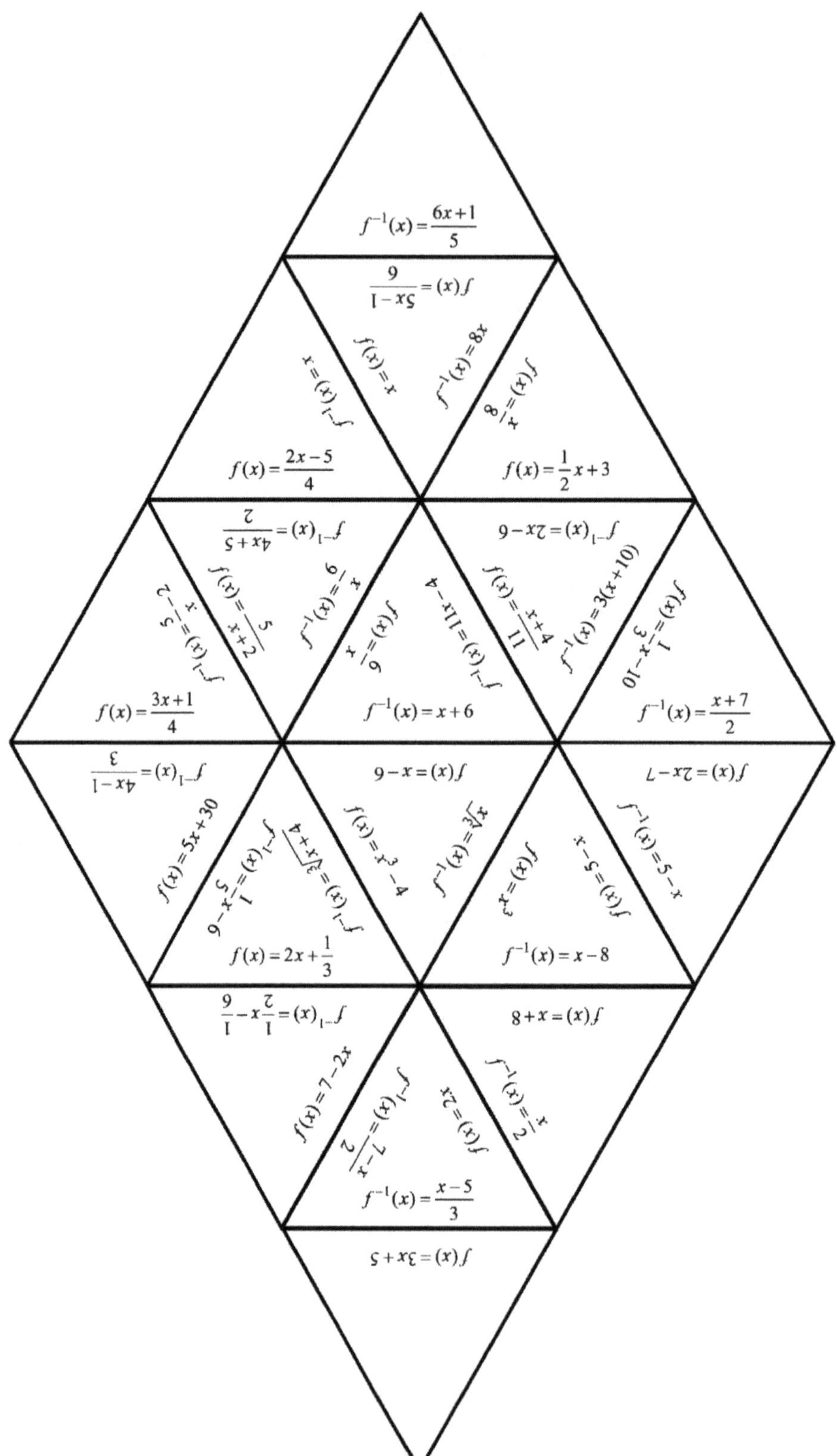

Answers for Alternative approach questions

1 Functions such as $f(x)=x$ or $f(x)=\frac{m}{x}$ where m is any number.

2 Functions such as $f(x)=x+2$ or $f(x)=x^3$ or $f(x)=3x$ because they only contain one step.

Inverse functions jigsaw pieces

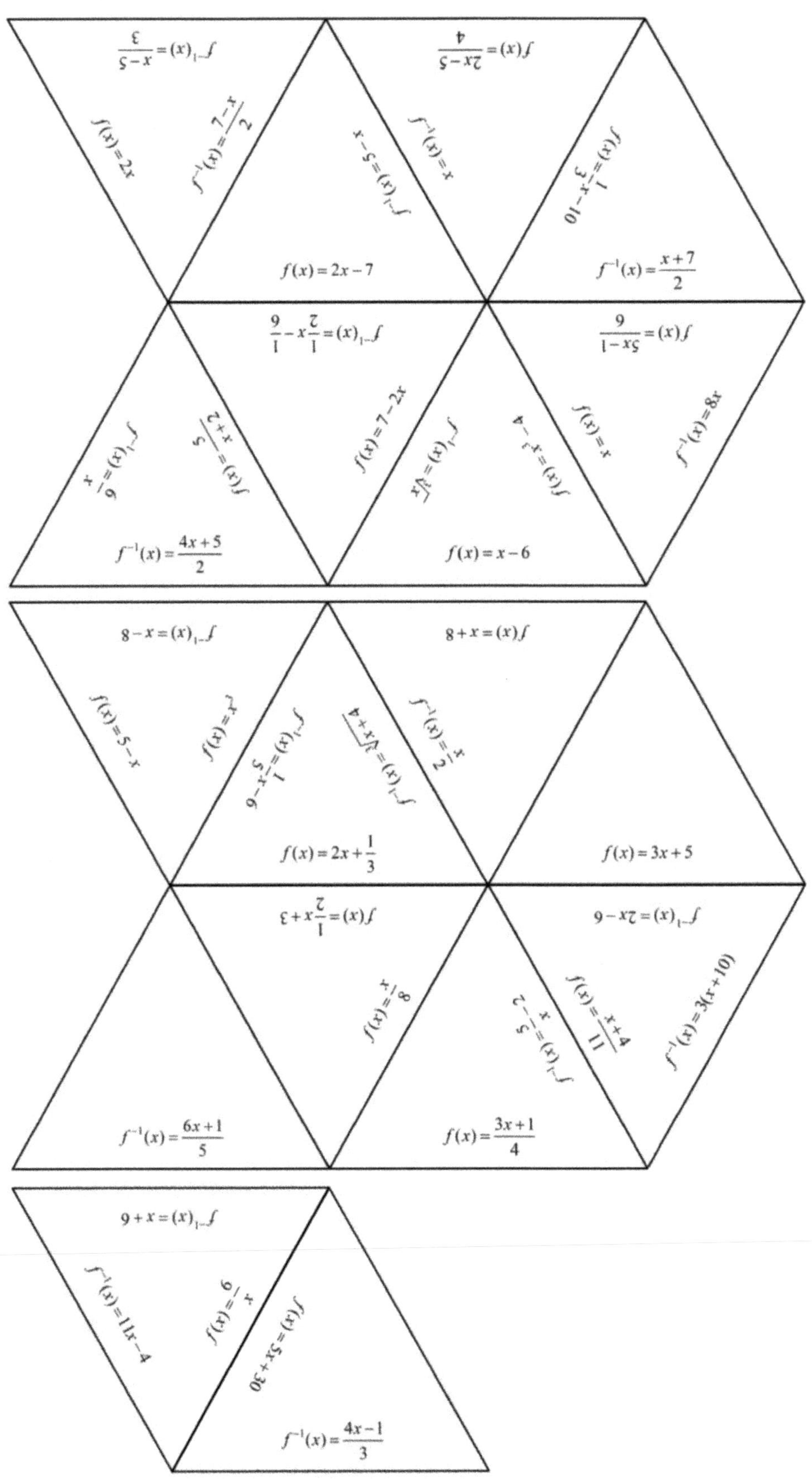

Larger size jigsaw pieces are available in the online resources.

Expanding or factorising jigsaw

Teacher notes

This jigsaw activity is an interesting group work alternative to a written exercise. The advantages of a jigsaw activity over a written exercise are that it creates opportunities for students to discuss their ideas, and they can find an alternative starting point if they are not sure how to do a particular question. Students are able to start with the answer and work backwards, which is sometimes an interesting way to tackle a question.

This activity allows students to use the same jigsaw to practise expanding or factorising. The majority of the expressions in this activity have double brackets, each with two terms. There are others such as single brackets, double brackets with more than two terms or expressions requiring expanding and then simplifying.

Introductory activity

If you want to use this task as a revision activity or a plenary you may find that little introduction is necessary. Otherwise you may want to go through the following examples.

Factorise $x^2 + 5x = x(x + 5)$

Factorise $x^2 + 3x - 10 = (x - 2)(x + 5)$

Expand and simplify

$$(4x - 3)(2x - 5)$$
$$= 8x^2 - 20x - 6x + 15$$
$$= 8x^2 - 26x + 15$$

Expand and simplify

$$(x^2 + 3x - 2)(x + 5)$$
$$= x^3 + 5x^2 + 3x^2 + 15x - 2x - 10$$
$$= x^3 + 8x^2 + 13x - 10$$

Factorise $2x^2 + 5x + 3 = (2x + 3)(x + 1)$

Expand and simplify

$$3(x + 2) - 2(x - 4)$$
$$= 3x + 6 - 2x + 8$$
$$= x + 14$$

Jigsaw activity

Ask your students to work in small groups. Copy the three sub-sets of jigsaw pieces to make a complete set of jigsaw pieces for each group. There are several different-shaped jigsaws in the Teacher Resource Kit so you should not tell the students what shape to expect. This jigsaw does not have any blank edges and this means that it is harder for your students to work out where the edges are.

Students can expand and then simplify the algebraic expressions or they may factorise.

Differentiation

To make it more difficult there are questions or 'solutions' associateed with common misconceptions around the outside edges of the jigsaw. If you want to make it easier, you could indicate outside edges; perhaps draw a bold line along the outside edges.

Alternative approaches

You could just ask students to complete the jigsaw or you could make use of the fact that the outside edges are not blank. You could copy blank jigsaw pieces from the online resources and ask students to use these pieces to extend the jigsaw, either finding the solutions to the harder questions, or writing a question that has the given answer. Alternatively students could make their own entirely new jigsaw.

Answers

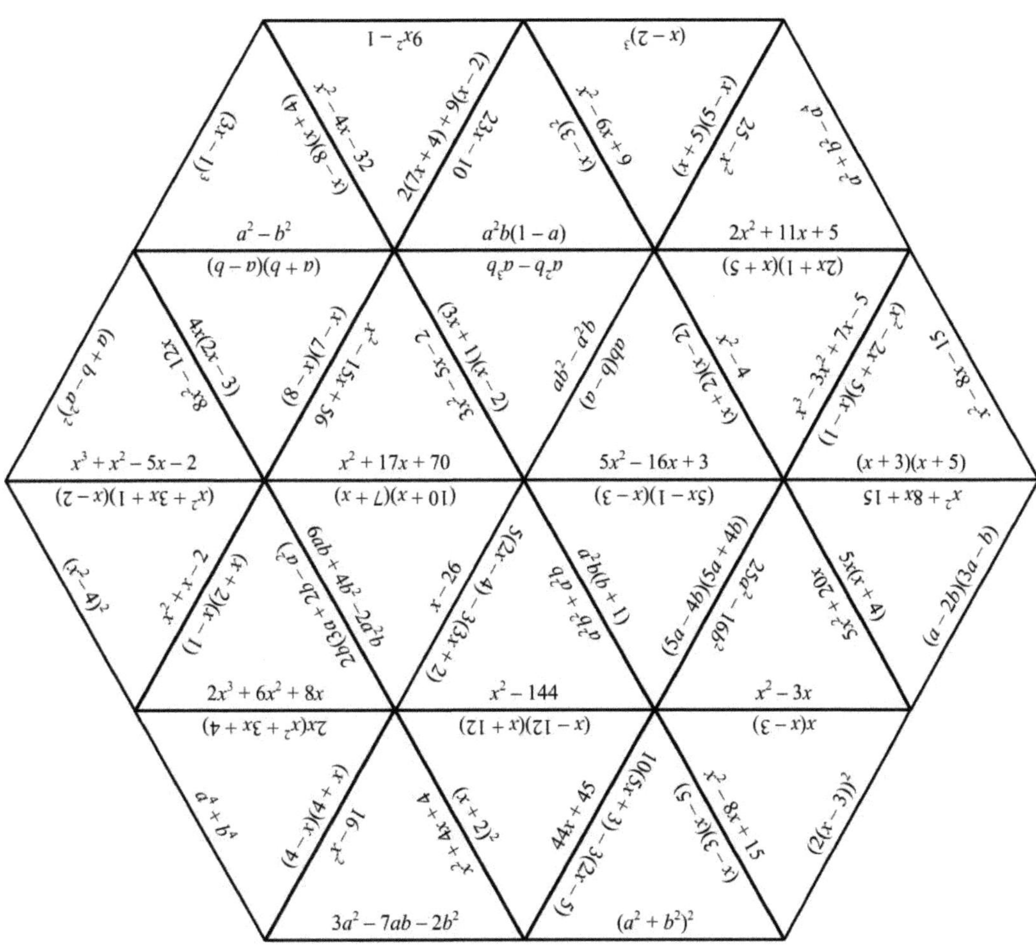

Expanding or factorising jigsaw pieces

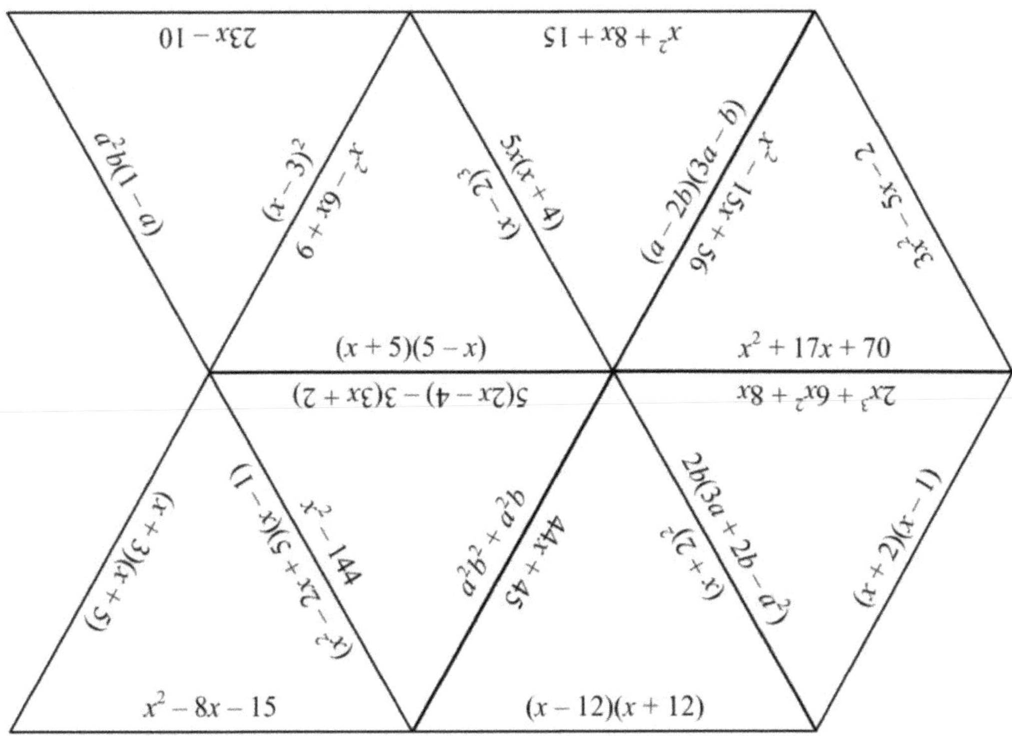

Expanding or factorising jigsaw pieces – continued

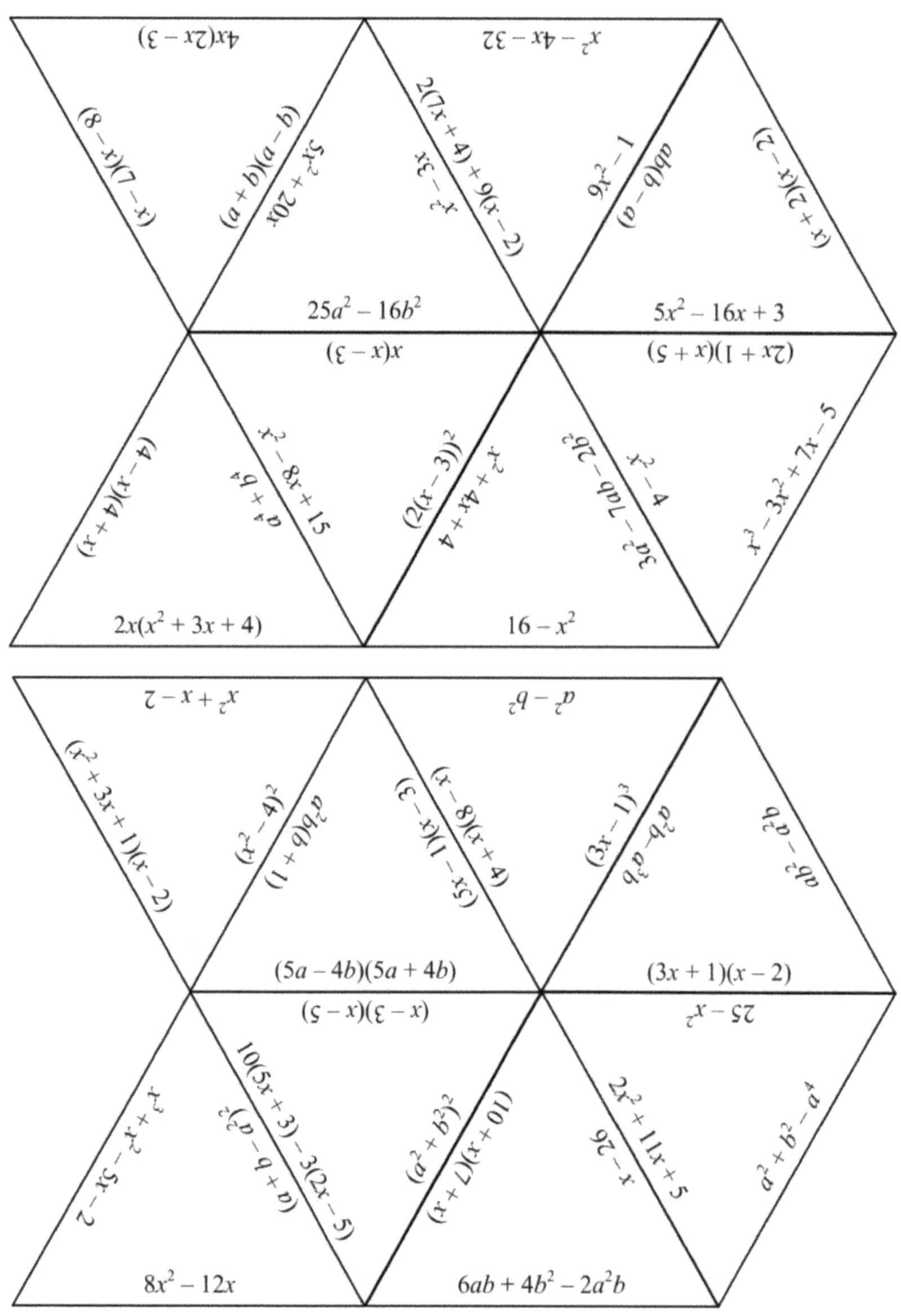

Larger size jigsaw pieces are available in the online resources.

Surface area and volume carousel

Teacher notes

This activity is a revision activity practising exam-style questions on surface area and volume. It is a good way to engage students and encourage group work. Students get to move around the classroom between questions, which is a good opportunity to increase their alertness and attention and reduce anxiety. Students also need to be confident at

- rounding to three significant figures
- using the area and arc length of a sector formulae
- using the density = mass ÷ volume formula
- using Pythagoras to work out the slant height of a cone or pyramid
- using the π key on the calculator.

Introductory activity

You may want to provide students with these formulae which would be given to them in an exam. SA = surface area, V = volume.

Cone h = perpendicular height, r = radius and l = slant height

$$\text{SA} = \pi r^2 + \pi r l \qquad \text{V} = \frac{1}{3}\pi r^2 h$$

Sphere r = radius $\quad \text{SA} = 4\pi r^2 \quad \text{V} = \frac{4}{3}\pi r^3$

Pyramid $\text{V} = \frac{1}{3} \times \text{base area} \times \text{height}$

Carousel activity

Set up six different desks in the classroom, each with a different exam-style question on it. Ask students to work in small groups and for each group to pick one of the desks to go to. Give each group an answer sheet, which can just be a plain piece of paper. When each group has chosen their first desk they have five minutes to do the question on that desk (you can find a timer at http://www.online-stopwatch.com/). When the five minutes are up the students move on to the next desk to do the next question for another five minutes. Repeat this until they have done all six questions. Allow time at the end of the lesson for providing the answers to the questions and for working through solutions as necessary. Inform students that all answers should be rounded to three significant figures.

Differentiation

You could make it easier or harder by:

- adjusting the time students have for the questions
- choosing which questions you use, you do not need to use all of these or can add to them by taking questions from past papers.

There are also two blank cards where students can design their own questions to do with surface area and volume. They should also include the full working with the answer. This can then be shared with other members of the class.

Alternative approaches

You could ask students to work individually instead of in groups. If you don't want students moving around the classroom you can give the questions to each group but still maintain the five minute rule for answering them. You could turn the questions into "show that…." questions so that students need to present full, clear and correct working.

Answers

1 a $\dfrac{110 \div 3}{360} \times 100 = 10.2\%$

b $110 : 250 = 11 : 25$

c $\dfrac{360 - 110}{360} \times \pi \times \left(\dfrac{25}{2}\right)^2 \times 10 = 3410 \text{ cm}^2$

2 a $\dfrac{4}{3} \times \pi \times 4.5^3 \div \pi \times \left(\dfrac{15}{2}\right)^2 = 2.16 \text{ cm}$

b $\sqrt[3]{\dfrac{800}{13} \times \dfrac{3}{4\pi}} = 2.45 \text{ cm}$

3 a $\dfrac{2}{3} \times \pi \times 15^3 \div \left(\pi \times \left(7^2 - 4.5^2\right)\right) = 78.3 \text{ cm}$

b $2 \times \pi \times 15^2 + \pi \times 15^2 = 2120 \text{ cm}^2$

4 a $\left(\pi \times 1.2^2 \times 6 + \dfrac{1}{3} \times \pi \times 1.2^2 \times 3.6\right) \times 10.5$

$= 342 \text{ g}$

b $\pi \times 1.2^2 + \pi \times 2.4 \times 6 + \pi \times 1.2 \times$

$\sqrt{3.6^2 + 1.2^2} = 64.1 \text{ cm}^2$

5 a $\dfrac{2}{3}$

b $\dfrac{1}{3} \times 7^2 \times 12 = 196 \text{ cm}^3$

c $7^2 + 4 \times \dfrac{1}{2} \times 7 \times \sqrt{3.5^2 + 12^2} = 224 \text{ cm}^2$

6 $7.6 \times 8 \times 2 + 2 \times \dfrac{25}{360} \times \pi \times 8^2 +$

$7.6 \times \dfrac{25}{360} \times \pi \times 8 \times 2 = 176.0 \text{ cm}^2$

$\sqrt{\left(\dfrac{176.0 - 2 \times \pi \times 3.2^2}{\pi \times 3.2}\right)^2 - 3.2^2} = 10.6 \text{ cm}$

Question 1

A circular cake has a diameter of 25 cm and a height of 10 cm.
A sector with an angle of 110° is cut from it and eaten.

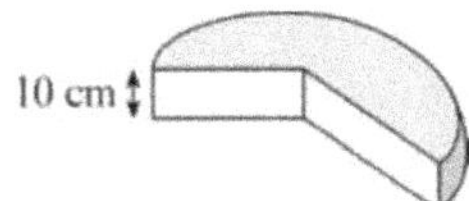

- **a** The cake that is eaten is shared between three people. Calculate the percentage of the whole cake that each person receives.
- **b** Write the ratio volume of cake eaten to volume of cake remaining in its simplest form.
- **c** Calculate the volume of the cake that remains.

Question 2

The diagram shows a cylinder with a diameter of 15 cm.
It contains water to a depth of 9 cm.
A metal sphere with radius 3.5 cm is placed in the water.

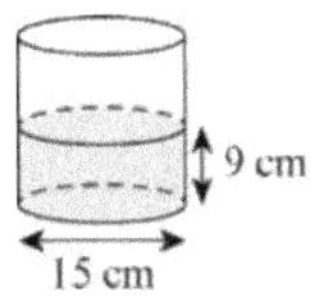

- **a** Calculate the increase in the height of the water.
- **b** A different metal sphere has a mass of 0.8 kg. One cubic centimetre of the metal has a mass of 13 grams. Work out the radius of this sphere.

Question 3

A metal hemisphere has a radius of 15 cm.
A metal pipe has an inner radius of 4.5 cm and an outer radius of 7 cm.
The hemisphere and the pipe have the same volume of metal in them.

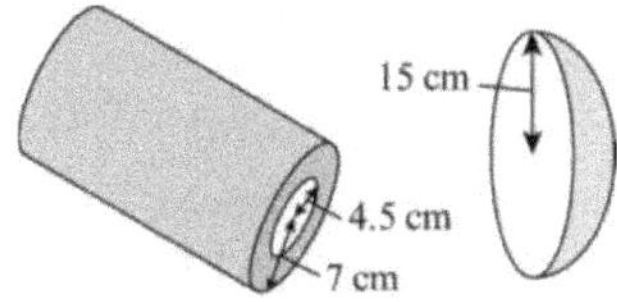

- **a** Work out the length of the pipe.
- **b** Work out the surface area of the hemisphere.

Question 4

Silver is used to make a solid.
Silver has a density of 10.5 g/cm^3.
The solid is made from a cone on top of a cylinder with the same radius, 1.2 cm.
The cylinder has a height of 6 cm and the cone has a height of 3.6 cm.

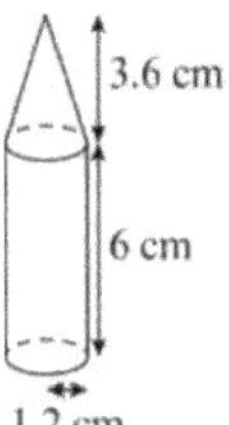

- **a** Find the mass of the solid.
- **b** Find the surface area of the solid.

Question 5

A solid square based pyramid fits exactly inside a box in the shape of a cuboid.
C is the centre of the base of the pyramid.

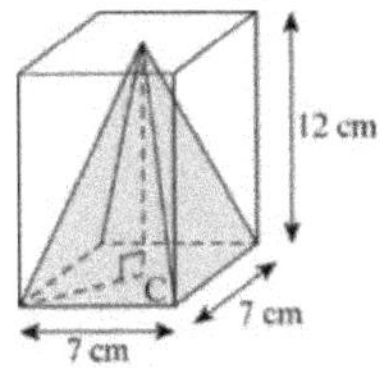

- **a** Find the fraction of the box not occupied by the solid.
- **b** Work out the volume of the solid.
- **c** Find the surface area of the solid.

Question 6

The diagram shows two solids with the same surface area.
The first solid is a prism of height 7.6 cm with cross section in the shape of a sector with radius 8 cm and sector angle 25°.
The second solid is made up from a hemisphere and a cone joined together. They both have the same radius of 3.2 cm.

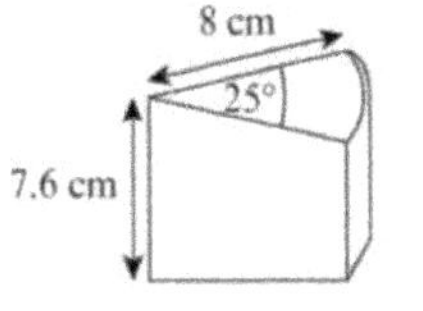

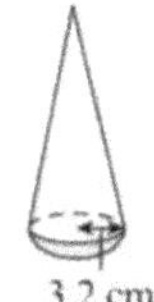

Work out the height of the cone.

Question 7

Design your own question to do with surface area or volume here. Write the full solution and answer on the back.

Question 8

Design your own question to do with surface area or volume here. Write the full solution and answer on the back.

Larger size question cards, and more blank cards, are available in the online resources.

Similar shapes true or false activity

Teacher notes

Students should already be able to calculate area, perimeter and volume for a variety of shapes. This activity can be used as a discussion activity before beginning work on areas and volumes of similar shapes. The task will provide the opportunity for students to discuss whether a variety of statements about area, perimeter and volume are true, before considering how to use scale factors to calculate areas or volumes of similar shapes. This activity is about changing lengths on 2D and 3D shapes to consider what effect this has on perimeter, area, surface area and volume.

Introductory activity

A brief recap on area, perimeter and volume calculations may be necessary before beginning this topic as many students get area and perimeter mixed up, even at extension level. Explain to students that this activity is about deciding what effect changing the lengths in a 2D or 3D shape has on its perimeter, area, surface area and volume.

True or false activity

Ask students to work together in small groups. Print out one copy of the statement cards for each group and cut them up. Ask students to sort them into two piles: true statements and false statements.

Differentiation

Once the students have sorted the statements ask them to correct the false statements. You could add some extra statements of your own or ask students to write their own statements using blank cards.

Alternative approaches

You can do this activity as a class discussion with the true and false facts photocopied, enlarged and stuck on the board.

Answers

True statements
If you double the lengths of the sides of a rectangle the perimeter doubles.
If you multiply each length of the sides of a triangle by 3 the perimeter should be multiplied by 3.
If you multiply each length of the sides of a rectangle by 4 the area should be multiplied by 16.
If you halve all the side lengths of a pentagon then the perimeter halves.
If you halve the lengths of all the edges of a cuboid the volume will be $\frac{1}{8}$ of its original volume.

False statements	Corrected statements
If you double the lengths of the sides of a rectangle the area doubles.	If you double the lengths of the sides of a rectangle the area is multiplied by 4.
If you multiply each length of the sides of a triangle by 3 the area should be multiplied by 3.	If you multiply each length of the sides of a triangle by 3 the area is multiplied by 9.
If you double the diameter of a circle the circumference should be multiplied by 4.	If you double the diameter of a circle the circumference is doubled.
If you multiply the radius of a circle by 5 the area should be multiplied by 5.	If you multiply the radius of a circle by 5 the area is multiplied by 25.
If you double the length of all of the edges of a cuboid the surface area doubles.	If you double the length of all of the edges of a cuboid the surface area is multiplied by 4.
If you double the length of all of the edges of a cube the volume should be multiplied by 4.	If you double the length of all of the edges of a cube the volume is multiplied by 8.
If you halve all the side lengths of a hexagon then the area halves.	If you halve all the side lengths of a hexagon then the area is quartered.

Similar shapes true or false activity cards

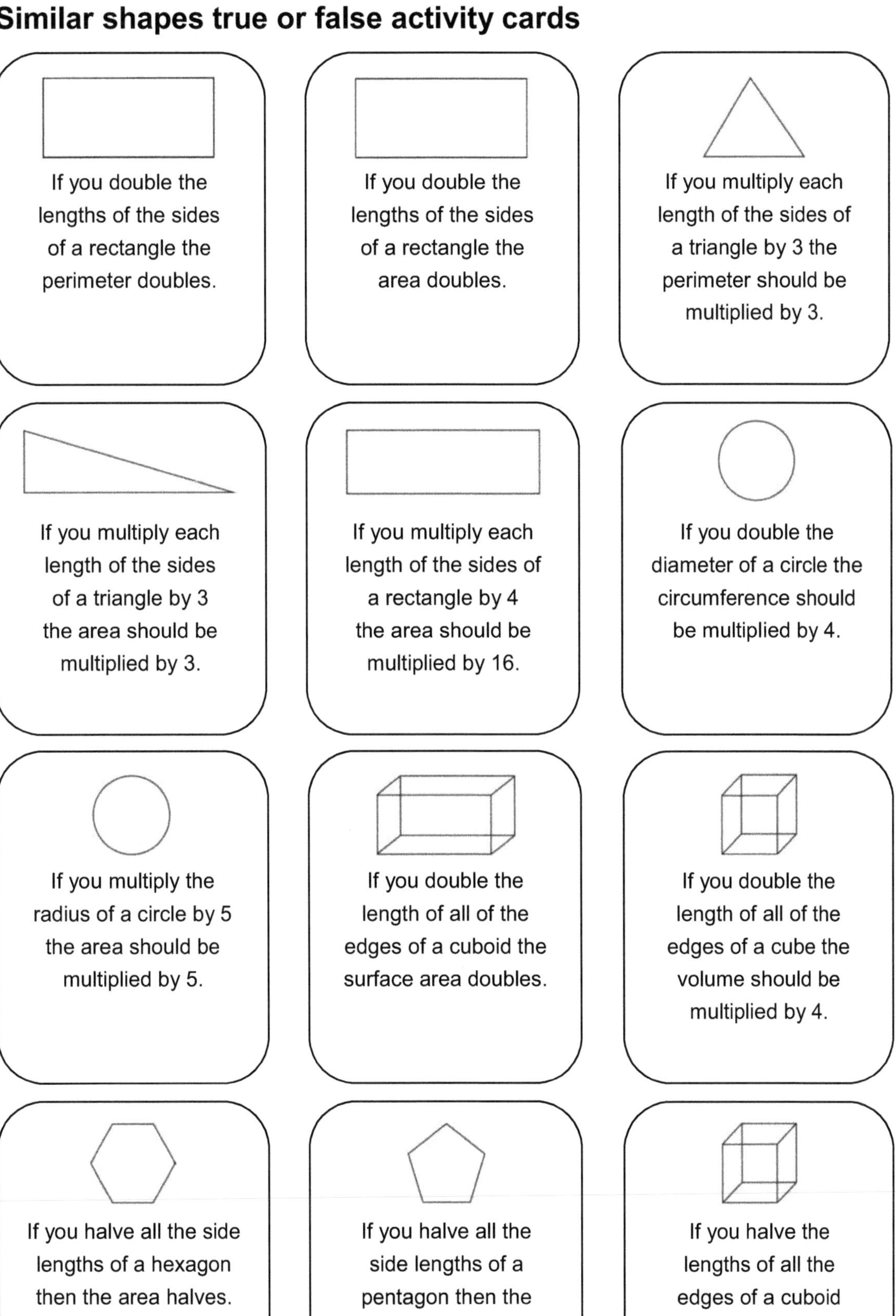
If you double the lengths of the sides of a rectangle the perimeter doubles.
If you double the lengths of the sides of a rectangle the area doubles.
If you multiply each length of the sides of a triangle by 3 the perimeter should be multiplied by 3.
If you multiply each length of the sides of a triangle by 3 the area should be multiplied by 3.
If you multiply each length of the sides of a rectangle by 4 the area should be multiplied by 16.
If you double the diameter of a circle the circumference should be multiplied by 4.
If you multiply the radius of a circle by 5 the area should be multiplied by 5.
If you double the length of all of the edges of a cuboid the surface area doubles.
If you double the length of all of the edges of a cube the volume should be multiplied by 4.
If you halve all the side lengths of a hexagon then the area halves.
If you halve all the side lengths of a pentagon then the perimeter halves.
If you halve the lengths of all the edges of a cuboid the volume will be $\frac{1}{8}$ of its original volume.

Speed, distance and time revision wheel

Teacher notes

When working on speed, distance and time one of the biggest problems for students is the fact that time uses bases 24 and 60 and not base 10. Before attempting to complete a calculation, students need to be confident when converting time into fractions or decimals, particularly if there are mixed units. It is very common to see 5 minutes 12 seconds written as 5.12 minutes instead of 5.2 minutes. Some of this activity is revising time conversions.

This activity is a jigsaw-type activity with only one solution. The activity revises the knowledge of the various rearrangements of the formula: speed = distance ÷ time, and there are also calculations of speed, distance and time to complete, some of which involve miss-matched units.

This is an interesting, alternative approach to using a written exercise as it promotes discussion, allows for group work, and the students often find it enjoyable. Students can do the questions in any order and can leave questions that they find difficult until the end. You and the students will know when they have finished as the puzzle will make a complete circle.

Introductory activity

This activity is intended as a revision activity, which may be completed at any point after having done the work on speed, distance and time in the student book. You may wish to give students no help at all or you could give the following examples as an introduction.

Example 1

Write 24 minutes as a fraction of an hour and as a decimal fraction of an hour.

24 minutes as a fraction of an hour is $\frac{24}{60} = \frac{2}{5}$ and as a decimal fraction of an hour is $24 \div 60 = 0.4$

Example 2

Change 0.3 of an hour into minutes.

$0.3 \text{ hrs} = 0.3 \times 60 = 18$ minutes

Example 3

In 1 hour 36 minutes a train travels 150 km at a constant speed. Find the speed of the train in km/h.

$$\text{Speed} = \frac{\text{distance}}{\text{time}}$$

First convert the time into hours only,

$$\text{time} = 1 \text{ hr } 36 \text{ min} = 1\frac{36}{60} \text{ hrs},$$

$$\text{speed} = \frac{150}{1\frac{36}{60}} = 93.75 \text{ km/h}$$

Example 4

A bird flies for 20 minutes at a speed of 6 m/s. Find the distance, in km, travelled by the bird.

distance = speed × time
(watch out for miss-matched units)

time = 20 × 60 = 1200 s,
distance = 6 × 1200 = 7200 m = 7.2 km

Jigsaw activity

Print out copies of the circular jigsaw pieces and cut them up. There are 24 pieces in each set. Ask the students to work in small groups to complete the jigsaw. There is only one correct solution.

Differentiation

You could set a time limit for this activity, or challenge the students to see which group can complete the activity first. You could leave out two or three cards in a section and ask students to make their own cards in order to complete the wheel. Blank cards are available in the online resources.

Alternative approaches

You could enlarge the jigsaw pieces to do one large jigsaw as a class or ask students to be creative to complete their own version (blank jigsaw pieces are available in the online resources).

Answers

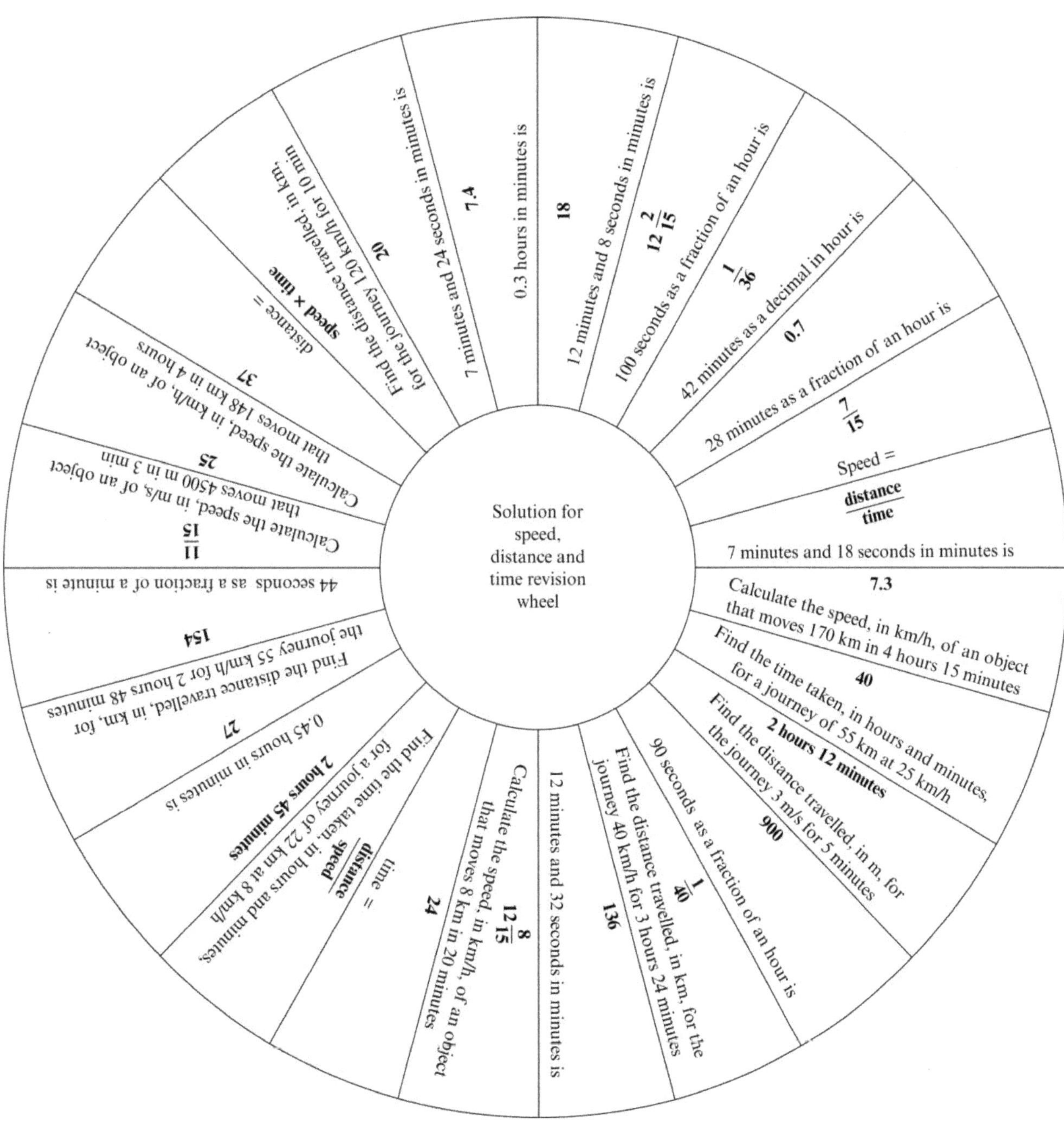

Solution for speed, distance and time revision wheel
0.3 hours in minutes is
18
12 minutes and 8 seconds in minutes is
$12\frac{2}{15}$
100 seconds as a fraction of an hour is
$\frac{1}{36}$
42 minutes as a decimal in hour is
0.7
28 minutes as a fraction of an hour is
$\frac{7}{15}$
Speed = $\frac{\text{distance}}{\text{time}}$
7 minutes and 18 seconds in minutes is
7.3
Calculate the speed, in km/h, of an object that moves 170 km in 4 hours 15 minutes
40
Find the time taken, in hours and minutes, for a journey of 55 km at 25 km/h
2 hours 12 minutes
Find the distance travelled, in m, for the journey 3 m/s for 5 minutes
900
90 seconds as a fraction of an hour is
$\frac{1}{40}$
Find the distance travelled, in km, for the journey 40 km/h for 3 hours 24 minutes
136
12 minutes and 32 seconds in minutes is
$12\frac{8}{15}$
Calculate the speed, in km/h, of an object that moves 8 km in 20 minutes
24
time = $\frac{\text{distance}}{\text{speed}}$
Find the time taken, in hours and minutes, for a journey of 22 km at 8 km/h
2 hours 45 minutes
0.45 hours in minutes is
27
Find the distance travelled, in km, for the journey 55 km/h for 2 hours 48 minutes
154
44 seconds as a fraction of a minute is
$\frac{11}{15}$
Calculate the speed, in m/s, of an object that moves 4500 m in 3 min
25
Calculate the speed, in km/h, of an object that moves 148 km in 4 hours
37
distance = speed × time
Find the distance travelled, in km, for the journey 120 km/h for 10 min
20
7 minutes and 24 seconds in minutes is
7.4

Speed, distance and time sort cards

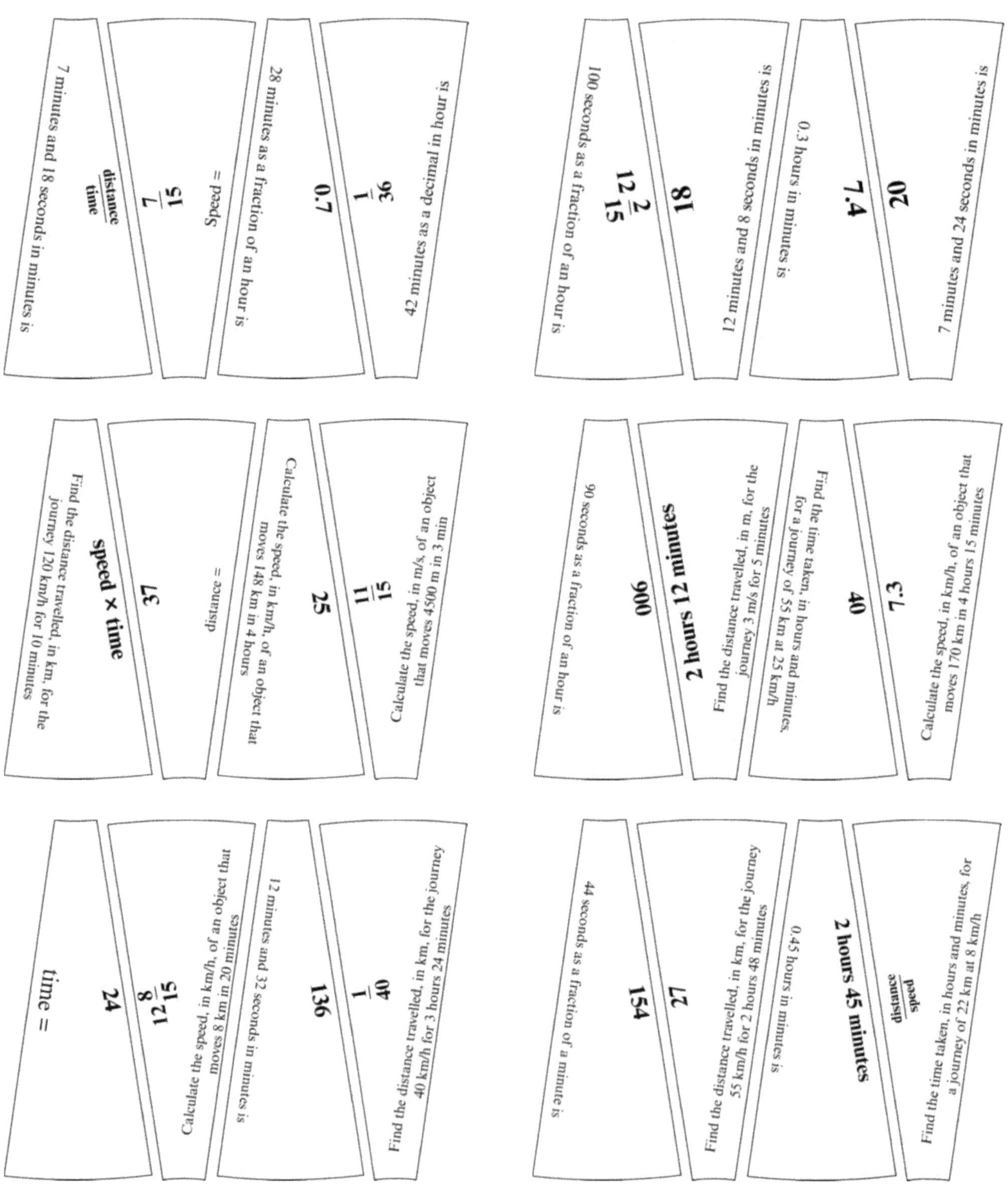
7 minutes and 24 seconds in minutes is
20
7.4
0.3 hours in minutes is
12 minutes and 8 seconds in minutes is
18
$12\frac{2}{15}$
100 seconds as a fraction of an hour is
Calculate the speed, in km/h, of an object that moves 170 km in 4 hours 15 minutes
7.3
40
Find the time taken, in hours and minutes, for a journey of 55 km at 25 km/h
Find the distance travelled, in m, for the journey 3 m/s for 5 minutes
2 hours 12 minutes
900
90 seconds as a fraction of an hour is
Find the time taken, in hours and minutes, for a journey of 22 km at 8 km/h
$\frac{\text{distance}}{\text{speed}}$
2 hours 45 minutes
0.45 hours in minutes is
Find the distance travelled, in km, for the journey 55 km/h for 2 hours 48 minutes
27
154
44 seconds as a fraction of a minute is
42 minutes as a decimal in hour is
$\frac{1}{36}$
0.7
28 minutes as a fraction of an hour is
Speed =
$\frac{7}{15}$
$\frac{\text{distance}}{\text{time}}$
7 minutes and 18 seconds in minutes is
Calculate the speed, in m/s, of an object that moves 4500 m in 3 min
$\frac{11}{15}$
25
Calculate the speed, in km/h, of an object that moves 148 km in 4 hours
distance =
37
speed × time
Find the distance travelled, in km, for the journey 120 km/h for 10 minutes
Find the distance travelled, in km, for the journey 40 km/h for 3 hours 24 minutes
$\frac{1}{40}$
136
12 minutes and 32 seconds in minutes is
Calculate the speed, in km/h, of an object that moves 8 km in 20 minutes
$12\frac{8}{15}$
24
time =

Larger size sort cards are available in the online resources.

Venn diagrams card sort

Teacher notes

Many students struggle with the union and intersection notation, and shading regions on Venn diagrams can be a problem. This card sorting activity practises these skills. It tests the understanding of symbols such as $\mathscr{E}$, $\varnothing$, $\cup$, $\cap$ and $'$ notation and various combinations of these, including the use of brackets. It can be used as a revision activity or an activity following the introduction of set notation and Venn diagrams.

Introductory activity

You can discuss how you show the union ($\cup$) and intersection ($\cap$) of two sets using $(A \cup B) \cap B$ is:

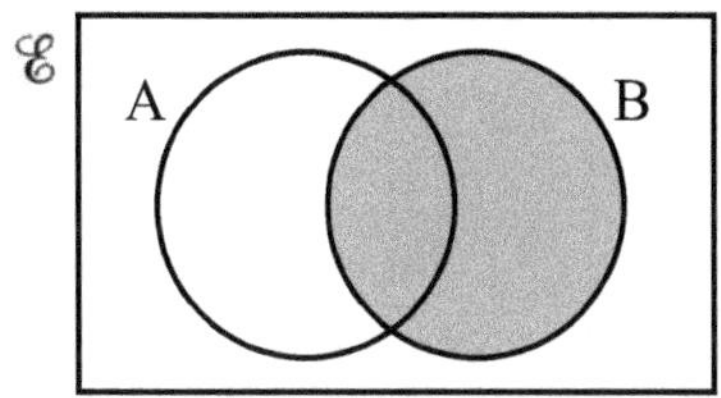

You could also discuss how this diagram would change if you introduced the 'not' ($'$) notation. Modifying the previous example $(A \cup B) \cap B'$ is:

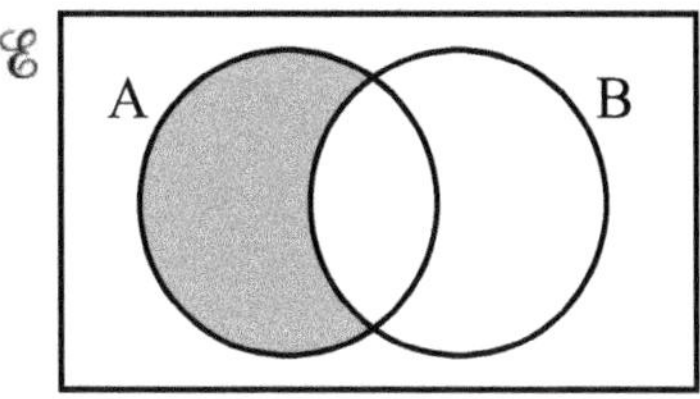

These two examples also use brackets, which may require further explanation.

Card sorting activity

Print out the sort cards and cut them up. Ask your students to work in small groups and sort the cards into matching sets. Each matched set should contain a diagram and matching notation card(s). The diagrams have between one and four matching cards and there is one set notation card that doesn't have a diagram to match it at all.

Differentiation

To make this activity a little easier you may chose to tell students how many set notation cards match each diagram (and that there is no diagram for one of the set notation cards). For the set notation card where a diagram does not exist, students could be asked to draw it; use one of the blank Venn diagram cards provided in the online resources.

You could extend this activity to look at three overlapping sets. Blank diagrams are provided in the online resources. Ask students to devise their own set of matching set notation cards for their own diagrams using these blank cards.

Alternative approaches

You could ask students to make their own set of diagrams and matching set notation cards, using the blank cards. There are a number of different diagrams that are not included in the initial card sorting activity.

Answers

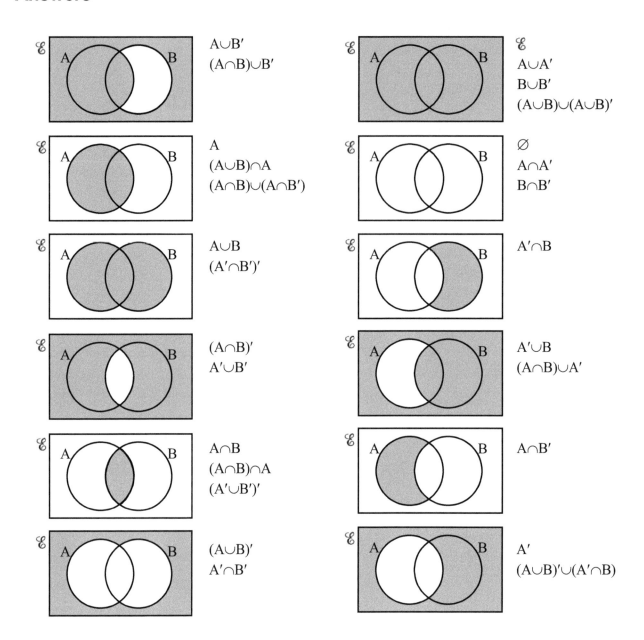

$A \cup B'$
$(A \cap B) \cup B'$

A
$(A \cup B) \cap A$
$(A \cap B) \cup (A \cap B')$

$A \cup B$
$(A' \cap B')'$

$(A \cap B)'$
$A' \cup B'$

$A \cap B$
$(A \cap B) \cap A$
$(A' \cup B')'$

$(A \cup B)'$
$A' \cap B'$

$\mathscr{E}$
$A \cup A'$
$B \cup B'$
$(A \cup B) \cup (A \cup B)'$

$\varnothing$
$A \cap A'$
$B \cap B'$

$A' \cap B$

$A' \cup B$
$(A \cap B) \cup A'$

$A \cap B'$

A'
$(A \cup B)' \cup (A' \cap B)$

No diagram for

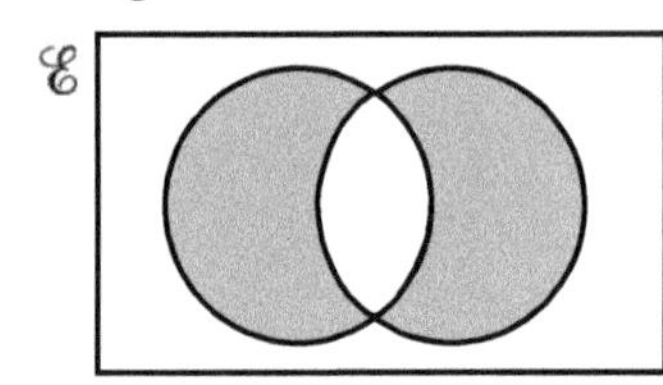

$(A \cap B)' \cap (A \cup B)$

Venn diagrams sort cards

$A\cap B'$	$(A\cup B)'$	$A'\cup B$
$(A\cap B)'$	A	$A\cap A'$
$A'\cap B$	$A\cup A'$	$A\cup B$
$A\cap B$	$\varnothing$	A'
$A\cup B'$	$\mathscr{E}$	$(A\cap B)\cup B'$
$(A'\cap B')'$	$A'\cap B'$	$A'\cup B'$
$(A\cap B)\cup A'$	$B\cup B'$	$(A'\cup B')'$
$B\cap B'$	$(A\cap B)\cap A$	$(A\cup B)\cap A$
$(A\cup B)\cup(A\cup B)'$	$(A\cap B)'\cap(A\cup B)$	$(A\cup B)'\cup(A'\cap B)$
		$(A\cap B)\cup(A\cap B')$

Larger size jigsaw pieces are available in the online resources.

Venn diagram sort cards – continued

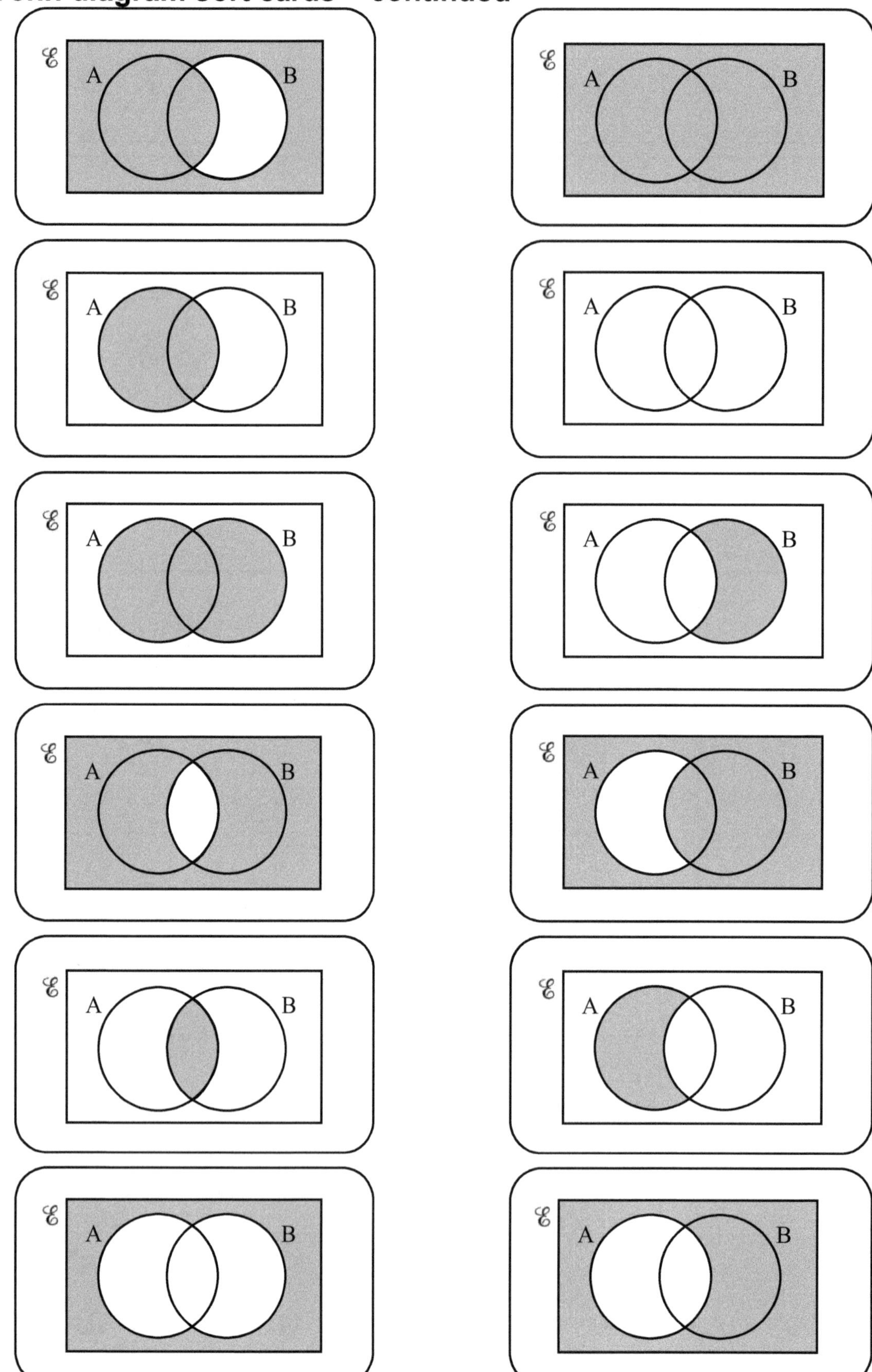

Larger size jigsaw pieces are available in the online resources.

Indices (including fractional) card sort

Teacher notes

The purpose of this indices card sort activity is to give students the opportunity to compare and contrast algebraic terms, using the laws of indices, in order to understand the different ways terms can be equivalent. The activity requires students to be able to use positive, negative, zero and fractional indices, and covers the laws of indices.

$x^a \times x^b = x^{a+b}$ $\quad x^a \div x^b = x^{a-b}$ $\quad (x^a)^b = x^{ab}$

$x^0 = 1$ $\quad x^{-a} = \frac{1}{x^a}$ $\quad x^{\frac{1}{b}} = \sqrt[b]{x}$

$x^{\frac{a}{b}} = \sqrt[b]{(x)^a}$ or $x^{\frac{a}{b}} = \sqrt[b]{(x)}^a$

In many cases a combination of these laws is needed.

Introductory activity

If you want to use this as a revision activity, or a plenary, you may find that little introduction is necessary. Otherwise you may want to revisit the laws of indices above.

Card sorting activity

Ask your students to work in small groups. You will need to copy and cut out the 33 cards to make one complete set for each group. Ask students to use the laws of indices to match cards into groups that are equivalent. The strength of this activity is that students do not know how many different sets of cards there are or how many cards there are in each set. The number of cards in each set actually ranges from two to six and there are two cards that do not have any matching pairs at all.

Differentiation

You could ask the students to add new cards to each set, particularly those sets that do not have many members. Blank cards are included in the online resources that you could use for this. Encourage students to find challenging examples.

Alternative approaches

You could give students a blank set of cards and ask them to devise their own completely different set(s) of expressions.

Answers

Group	Cards for Group
x	x^1, $\frac{x^{-3}}{x^{-4}}$, $\frac{x^{50}}{x^{49}}$, $(\sqrt{x})^2$
$x^{\frac{1}{2}}$	$\frac{x}{\sqrt{x}}$, $x^{\frac{1}{2}}$, $\sqrt{x}$, $\frac{x^3}{x^{\frac{5}{2}}}$, $x \times x^{-\frac{1}{2}}$
$x^{\frac{1}{5}}$	$x^{\frac{1}{5}}$, $\sqrt[5]{x}$
x^2	$(\sqrt{x})^4$, $\frac{1}{x^{-2}}$, $\frac{x^{\frac{5}{2}}}{x^{\frac{1}{2}}}$
x^{-1}	$\frac{\sqrt{x}}{x^{\frac{3}{2}}}$, $\frac{1}{x}$, x^{-1}, $\frac{x^4}{x^5}$
$x^{-\frac{1}{2}}$	$\frac{\sqrt{x}}{x}$, $\frac{1}{\sqrt{x}}$, $\frac{1}{x^{\frac{1}{2}}}$, $x^{-\frac{1}{2}}$, $\frac{x^2}{\sqrt{x^5}}$, $x^0 \div \sqrt{x}$
$x^{\frac{5}{2}}$	$x^2\sqrt{x}$, $\frac{x^3}{\sqrt{x}}$, $\sqrt{x^5}$, $(\sqrt{x})^5$, $x^{\frac{5}{2}}$
$x^{\frac{2}{5}}$	$(\sqrt[5]{x})^2$, $\sqrt[5]{x^2}$
x^5	$\frac{1}{x^{-5}}$
1	x^0

Indices (including fractional) sort cards

$x^{\frac{1}{5}}$	$\frac{\sqrt{x}}{x}$	$\frac{\sqrt{x}}{x^{\frac{3}{2}}}$	$\left(\sqrt{x}\right)^4$	$\frac{1}{x}$
$x^2\sqrt{x}$	$\left(\sqrt[5]{x}\right)^2$	$\frac{x^3}{\sqrt{x}}$	$\frac{x}{\sqrt{x}}$	x^1
x^{-1}	$\frac{1}{\sqrt{x}}$	$\frac{x^{-3}}{x^{-4}}$	$x^{\frac{1}{2}}$	$\frac{1}{x^{\frac{1}{2}}}$
$\sqrt{x}$	$\frac{x^4}{x^5}$	$\sqrt[5]{x}$	$\sqrt{x^5}$	$\sqrt[5]{x^2}$
$\left(\sqrt{x}\right)^5$	$\frac{x^3}{x^{\frac{5}{2}}}$	$x \times x^{-\frac{1}{2}}$	$\frac{1}{x^{-5}}$	$\frac{1}{x^{-2}}$
$\frac{x^{50}}{x^{49}}$	$\frac{x^{\frac{5}{2}}}{x^{\frac{1}{2}}}$	$x^{-\frac{1}{2}}$	$x^{\frac{5}{2}}$	$\left(\sqrt{x}\right)^2$
		$\frac{x^2}{\sqrt{x^5}}$	$x^0 \div \sqrt{x}$	x^0

Larger size jigsaw pieces are available in the online resources.

Angle properties pairs game

Teacher notes

This activity can be used as a starter, a plenary or a revision exercise. Note that not all of the angle property facts are included in this resource.

Introductory activity

Give the students an example of a pair of cards. On the first card will be the "title", on the second card will be the angle property and a diagram illustrating the property. Explain the rules of the pairs game.

An isosceles triangle …

has two equal angles.

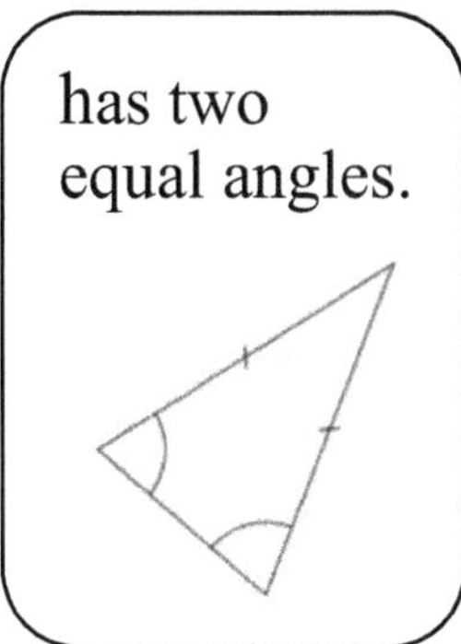

Pairs game

Ask students to work in pairs or small groups. Print out the cards and cut out, this makes one complete set of 12 pairs. You will need one set for each group of students. Ask the students to shuffle the cards and deal them face down in 4 rows of 6. Students take it in turns to turn two cards over, if they are a matching pair they keep the cards and have another go. If they are not a matching pair students must return them face down to the position that they came from. At the end of the game when all pairs are found the winner is the person with the most pairs.

Differentiation

Ask the students to come up with their own extra pair(s) using blank boxes. Since not all of the angle property facts are included in this resource students could try to identify which are missing.

Remove some cards so that there are less pairs to work with or make the task harder by adding extra cards of your own.

Alternative approaches

Students could play a game of snap with the cards.

Answers

Title	Property
Angles at a point …	add up to 360°.
Angles at a point on a straight line …	add up to 180°.
Intersecting straight lines …	form vertically opposite angles which are equal.
Angles formed within parallel lines …	alternate angles are equal, corresponding angles are equal.
Angles in the same segment of a circle …	are equal.
Cyclic quadrilaterals …	opposite angles add up to 180°.
Angles in a triangle …	add up to 180°.
Angles in a quadrilateral …	add up to 360°.
Exterior angles in a polygon …	add up to 360°.
Angle in a semi-circle …	is 90°.
Angle between tangent and radius of a circle …	is 90°.
Angle at the centre of a circle and angle at the circumference …	Angle at the centre of a circle is twice the angle at the circumference

Angle properties game cards

Angles at a point …	Angles at a point on a straight line …	Intersecting straight lines …
Angles formed within parallel lines …	Angles in the same segment of a circle …	Cyclic quadrilaterals …
Angles in a triangle …	Angles in a quadrilateral …	Exterior angles in a polygon …
Angle in a semi-circle …	Angle between tangent and radius of a circle …	Angle at the centre of a circle and angle at the circumference …

Angle properties game cards – continued

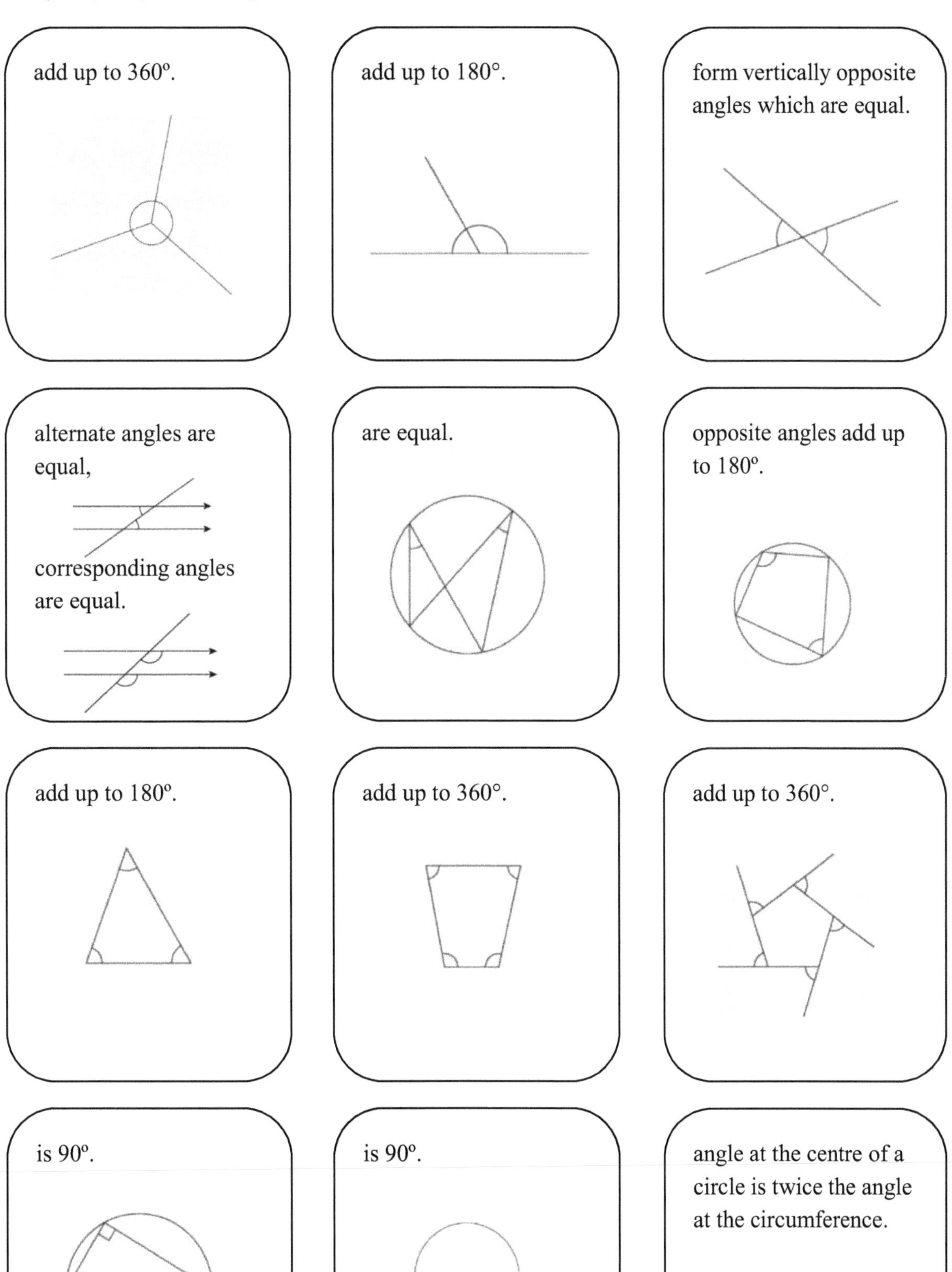

add up to 360°.
add up to 180°.
form vertically opposite angles which are equal.
alternate angles are equal,
corresponding angles are equal.
are equal.
opposite angles add up to 180°.
add up to 180°.
add up to 360°.
add up to 360°.
is 90°.
is 90°.
angle at the centre of a circle is twice the angle at the circumference.

Tree diagrams – three in a row game

Teacher notes

This three in a row game is for students to reinforce their learning about questions involving probability. It covers the work in unit 6 on combined and independent events and using tree diagrams to solve probability problems including examples with and without replacement.

Introductory activity

This is best used as a revision activity and requires no introduction if you use it immediately after studying the tree diagrams work in unit 6.

If you want to you can remind students of the rule P(A and B) = P(A) × P(B) for independent events. It might also be worth reminding students that when dealing with probabilities without replacement the denominators of the fractions need to reduce, and the numerators need be adjusted accordingly in tree diagrams, as it is very common for students to forget this.

3 in a row activity

Ask your students to work in pairs to play the game. The students need to take the pack of probability card questions and shuffle them. They also need a copy of the answer sheet and two different sets of 8 coloured counters.

The first student takes a question card from the shuffled deck. They answer the question and look for the correct answer on the answer sheet, when they find the correct answer they place their coloured counter on the answer sheet. The second student then takes their turn. Students continue to take turns to answer questions and place counters.

The winning student is the first student to place 3 counters in a row either horizontally, vertically or diagonally.

If a student can't find the answer on the answer sheet (because they have worked it out incorrectly) they do not place any counter and must return the question to the bottom of the shuffled cards and turn moves to the next student.

If the student makes a mistake in working out the answer and puts their counter on the wrong answer space the opponent may replace the counter with their own.

There is also a 'wild card'. If students get this card they can place their counter on the wild card space on the answer sheet without answering any questions and then the turn moves to the next student.

Differentiation

To make it easier you could partially draw the tree diagrams on some of the cards to start them off.

To extend the activity when the wild card is drawn, instead of it being a free go the student has to answer a question set by their opponent or by you.

Some of the cards are very easy, e.g. cards 9 and 16. You could replace these with harder questions.

Alternative approaches

If you do not have coloured counters students could also just write their names on the answer grid when they get an answer correct. Students could be asked to write their own probability questions to give the answers on the grid.

Question cards

Card 1

The spinner is divided into 5 equal sections. 2 are black and 3 are white.

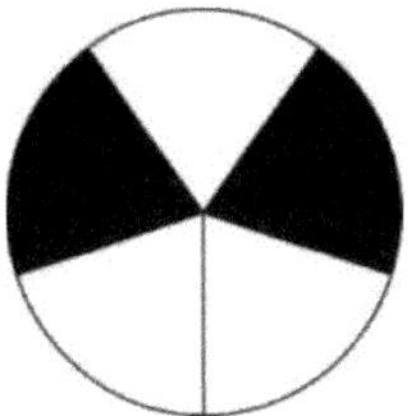

The spinner is spun twice. Find the probability that white is spun both times.

Card 2

Find the missing fraction on this tree diagram showing the probability of winning or losing two games. (Note the probability of drawing any game is 0).

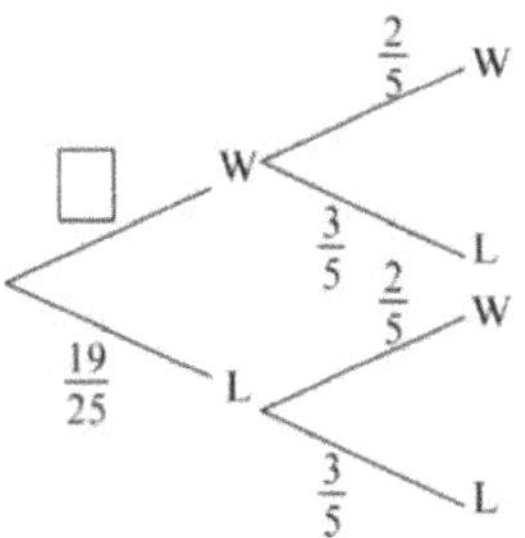

Card 3

The probability that an archer hits a target when she shoots an arrow is 0.8

Find the probability that she misses the target on her first two shots.

Card 4

A

B

Bag A contains 3 red counters and 2 blue counters. Bag B contains 7 red counters and 3 blue counters.

One counter is picked from each bag at random.

Find the probability of picking one counter of each colour.

Card 5

An unbiased six-sided dice is rolled twice.

Work out the probability that the first roll is a 6 and the second roll is an even number.

Card 6

Three unbiased coins are thrown.

Find the probability that all three throws are heads.

Card 7

The probability that a spinner lands on a 1 is 0.7

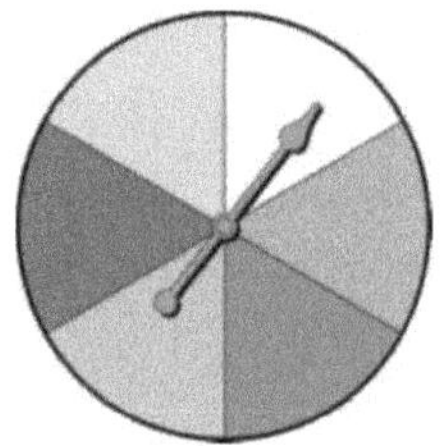

The spinner is spun three times.

Work out the probability that less than two of the spins shows a 1.

Card 8

Two letters are picked from the word MATHS at random. A tree diagram is drawn to show the probability of picking a T or not picking a T. Find the value of the missing probability x.

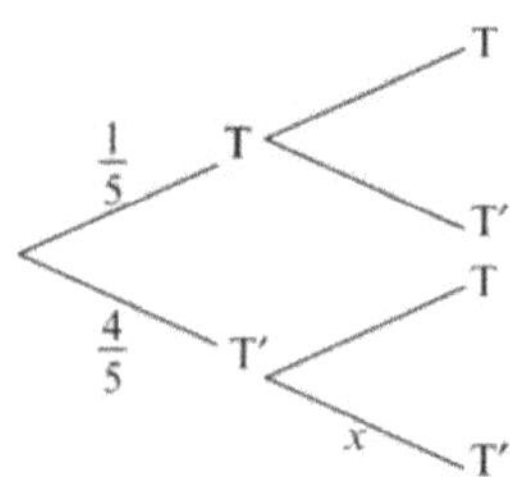

Card 9

The probability that it will rain tomorrow is 0.2

Work out the probability it will not rain tomorrow.

Card 10

Amy has a bag of 10 sweets. 7 are red and 3 are yellow.

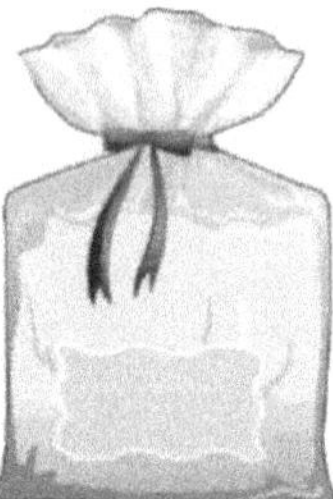

Amy eats two sweets at random. Work out the probability they are both red.

Card 11

Box A contains 1 green pen and 4 black pens.
Box B contains 1 green pen and 9 black pens.

One pen is picked from each box at random.

Find the probability of picking at least one green pen.

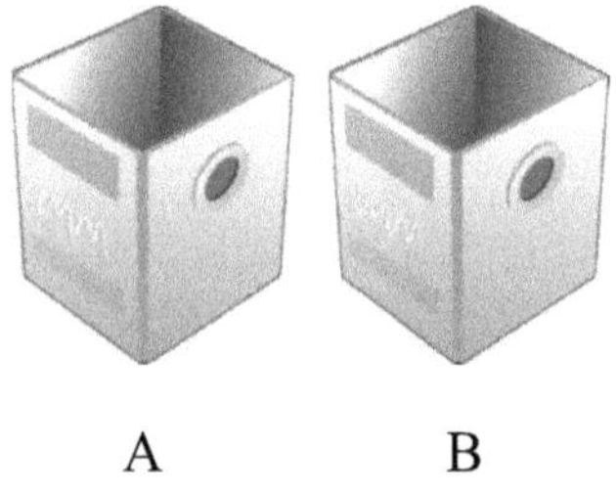

Card 12

The probability that Katy hits a 20 in a game of darts is 0.2.

Work out the probability that Katy hits the 20 for the first time on her fourth attempt.

Card 13

The probability of a spinner landing on a number 2 is 0.4.

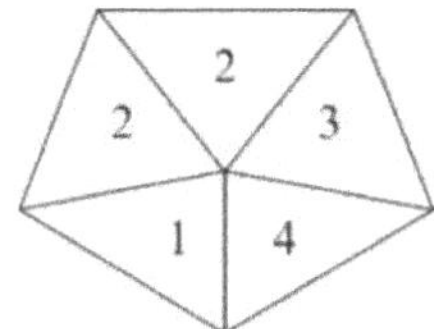

Work out the probability that the spinner lands on the number 2 for the first time on the fifth spin.

Card 14

An unbiased six-sided dice is rolled twice.

Work out the probability that neither roll is a 5.

Card 15

Three unbiased coins are thrown.

Find the probability that more than one throw is a head.

Card 16

The probability that Jamil goes to school by bike is 0.07

Work out the probability that Jamil does not go to school by bike.

Card 17

A bag contains 4 blue marbles, 5 green marbles and 3 white marbles.

Ami picks a marble from the bag at random, it is not replaced. She then picks a second marble.

Work out the probability that she picks two marbles of the same colour.

Card 18

Edward either gets the bus or walks to school. The probability that he gets the bus is 0.8. If he gets the bus to school the probability he is late is 0.1. If he walks to school the probability he is late is 0.3.

Find the probability that Edward is not late to school.

Card 19

A box contains 3 strawberry cakes and 2 lemon cakes. Katia picks two cakes at random from the box.

Find the missing probability x on the tree diagram.

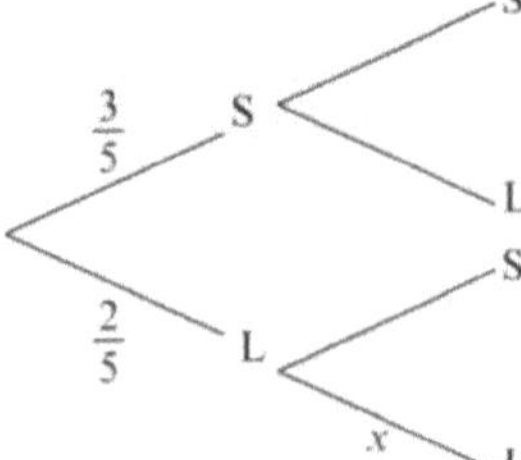

Card 20

A letter is picked from random from the 10 letters in the word STATISTICS. The letter is not replaced. A second letter is then picked at random. Find the probability that at least one of the letters is the letter S.

Card 21

An unbiased six-sided dice is rolled three times.

Work out the probability that exactly 2 fours are rolled.

Card 22

A bag contains 5 black counters and 3 white counters.

Alex picks a counter form the bag at random, it is not replaced. He then picks a second counter.

Work out the probability that he picks one black counter and one white counter.

Card 23

A letter is picked from random from the 7 letters in the word ALGEBRA. The letter is not replaced. A second letter is then picked at random. Find the probability that neither letter is the letter A.

Card 24

Hamish has 5 cards each with a shape on it.

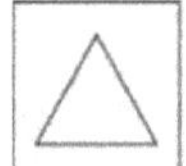 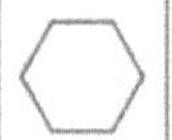 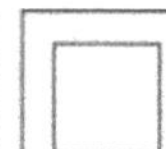  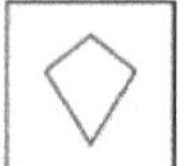

Hamish picks two cards at random without replacement. Find the probability that both cards picked have quadrilaterals on them.

Answer sheet

$\frac{9}{25}$	$\frac{3}{4}$	$\frac{7}{25}$	$\frac{1}{2}$	$\frac{8}{15}$
$\frac{19}{66}$	$\frac{1}{4}$	$\frac{6}{25}$	0.93	0.8
0.153	$\frac{5}{72}$	**WILD CARD**	0.04	$\frac{1}{10}$
0.1024	$\frac{23}{50}$	$\frac{7}{15}$	$\frac{10}{21}$	$\frac{1}{8}$
$\frac{25}{36}$	0.86	0.05184	$\frac{15}{28}$	$\frac{1}{12}$

Answers

Card 1 $\frac{3}{5}\times\frac{3}{5}=\frac{9}{25}$	Card 8 $\frac{3}{4}$	Card 11 $1-\frac{4}{5}\times\frac{9}{10}=\frac{7}{25}$	Card 15 $4\times\frac{1}{2}\times\frac{1}{2}\times\frac{1}{2}=\frac{1}{2}$	Card 20 $1-\frac{7}{10}\times\frac{6}{9}=\frac{8}{15}$
Card 17 $\frac{4}{12}\times\frac{3}{11}+\frac{5}{12}\times\frac{4}{11}+$ $\frac{3}{12}\times\frac{2}{11}=\frac{19}{66}$	Card 19 $\frac{1}{4}$	Card 2 $1-\frac{19}{25}=\frac{6}{25}$	Card 16 0.93	Card 9 0.8
Card 7 $2\times 0.3\times 0.3\times 0.7$ $+\ 0.3^3 = 0.153$	Card 21 $3\times\frac{1}{6}\times\frac{1}{6}\times\frac{5}{6}=\frac{5}{72}$	**WILD CARD**	Card 3 $0.2\times 0.2 = 0.04$	Card 24 $\frac{2}{5}\times\frac{1}{4}=\frac{1}{10}$
Card 12 $0.8^3\times 0.2 =$ 0.1024	Card 4 $\frac{3}{5}\times\frac{3}{10}+\frac{2}{5}\times\frac{7}{10}=\frac{23}{50}$	Card 10 $\frac{7}{10}\times\frac{6}{9}=\frac{7}{15}$	Card 23 $\frac{5}{7}\times\frac{4}{6}=\frac{10}{21}$	Card 6 $\frac{1}{2}\times\frac{1}{2}\times\frac{1}{2}=\frac{1}{8}$
Card 14 $\frac{5}{6}\times\frac{5}{6}=\frac{25}{36}$	Card 18 $0.8\times 0.9\ +$ $0.2\times 0.7 = 0.86$	Card 13 $0.6^4\times 0.4 =$ 0.05184	Card 22 $\frac{5}{8}\times\frac{3}{7}+\frac{3}{8}\times\frac{5}{7}=\frac{15}{28}$	Card 5 $\frac{1}{6}\times\frac{1}{2}=\frac{1}{12}$

Distance–time and speed–time graphs card sort

Teacher notes

One of the hardest concepts in distance–time and speed–time graphs is to interpret the graphs. In particular some students have poor understanding of possible and impossible features in a distance–time graph or speed–time graph. For example, vertical lines are not possible in either. The purpose of this activity is to give students the opportunity to compare and contrast graphs so they are able to understand the meaning of them. Covered in this card sort are the following points.

- The gradient of a distance–time graph gives the speed. The steeper the graph the faster the object is travelling.
- In a distance–time graph if the line is straight the speed is constant, if it is curved then the object is accelerating or decelerating, depending on the direction of the curve.
- A horizontal line in a distance–time graph shows that an object is stationary.
- A vertical line in a distance–time graph is impossible as it would imply infinite speed.
- In a distance–time graph the line (or curve) goes from left to right – otherwise it would show an object travelling backward in time.
- The gradient of a speed–time graph gives the acceleration, positive gradient indicates acceleration and negative gradient indicates deceleration. The steeper the graph the faster the acceleration or deceleration.
- In a speed–time graph if the line is straight the acceleration is constant.
- A horizontal line in a speed–time graph shows an object travelling at constant speed.

Introductory activity

You may want to use this activity as a revision activity in which case you may choose to give no introduction. Discussion from this activity will hopefully lead students to all the bullet points listed in the teacher notes.

Card sorting activity

Print out the sort cards and cut them up; ask students to work in small groups. Give each group a set of these cards. To keep it simple you could just ask students to sort the cards into two piles: one pile of **possible** graphs and one pile of **impossible** graphs. If you want to extend the activity ask the students to think about the meanings of the possible graphs and the reason why the impossible graphs are impossible.

Differentiation

You can also differentiate by leaving some difficult cards out for less able students or by adding harder cards to extend the more able students. You can allow students a "**don't know**" pile. These can be sorted when you move around the class.

Alternative approaches

You could extend the task further, by asking students to write some examples of their own. The students have a choice to do easy examples or really difficult ones. You could give them the incentive to do a difficult one by telling them their card will be given to another group for the other group to try to classify it. Use the blank cards from the online resources.

Answers

POSSIBLE		IMPOSSIBLE	
Graph with explanation of graph		**Graph with reason why it is impossible**	
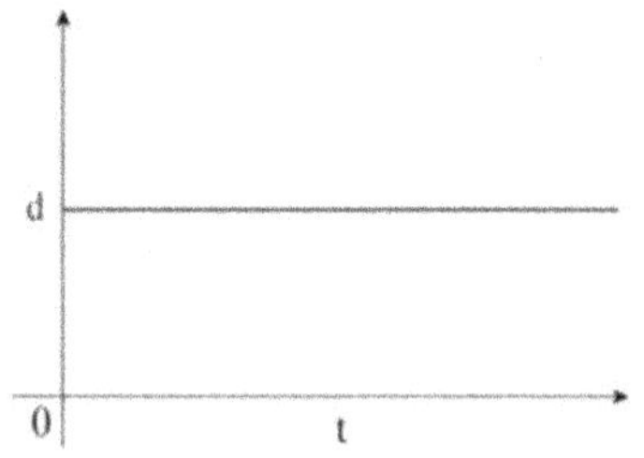	This shows a stationary object.	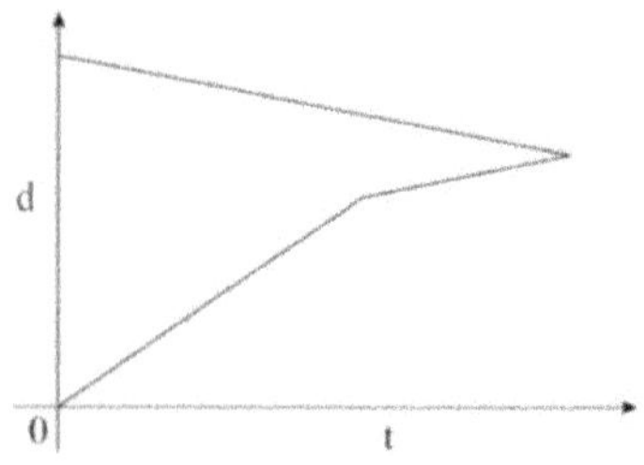	This graph is impossible as it shows going back in time.
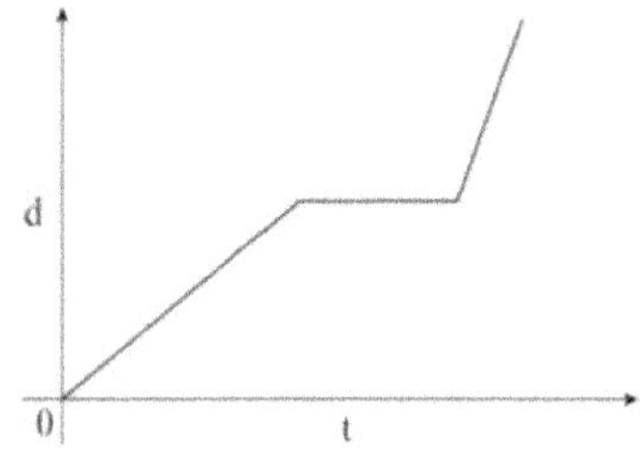	This object travels at constant speed, then stops, then travels at a faster speed.	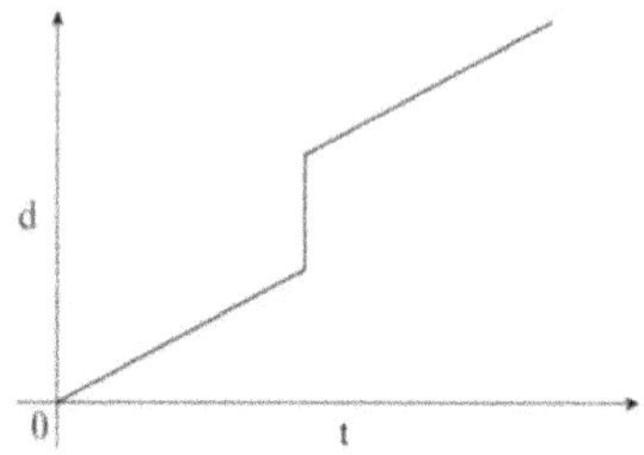	This graph is impossible as the vertical part shows infinite speed.
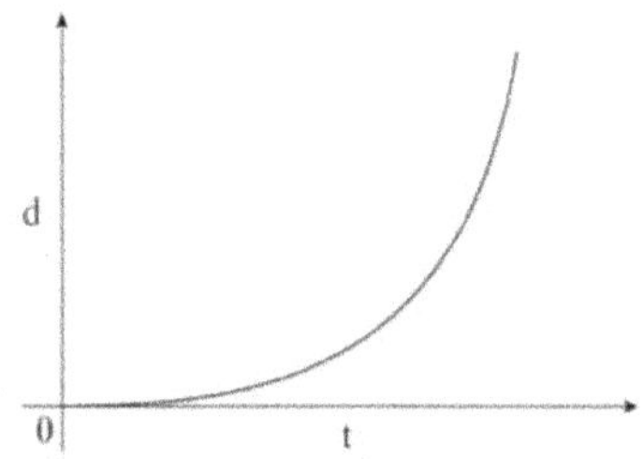	This object is accelerating.	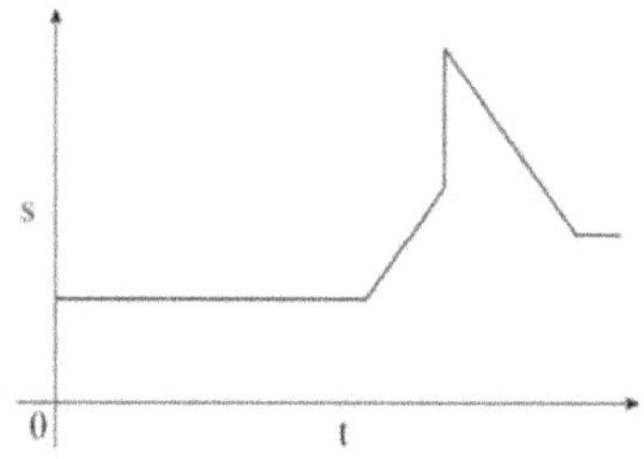	This graph is impossible as the vertical part shows infinite acceleration.
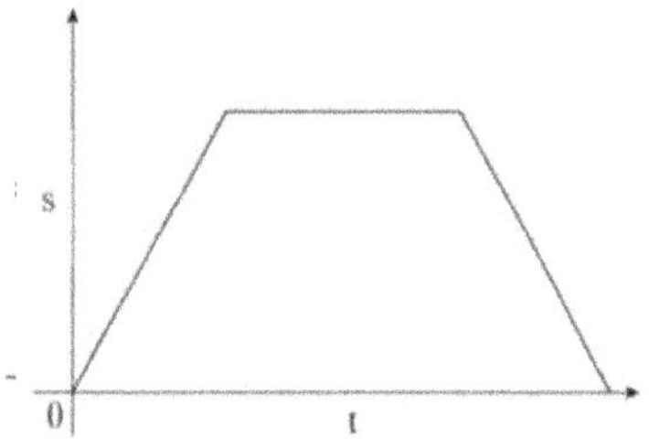	This object is accelerating at a constant rate, then travelling at constant speed then decelerating at a constant rate.		
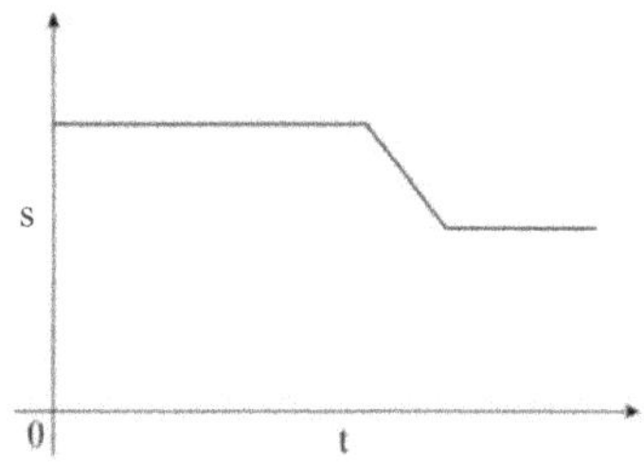	This object is travelling at constant speed, then decelerating at a constant rate, then travelling at a slower constant speed.		

Distance–time and speed–time graphs sort cards

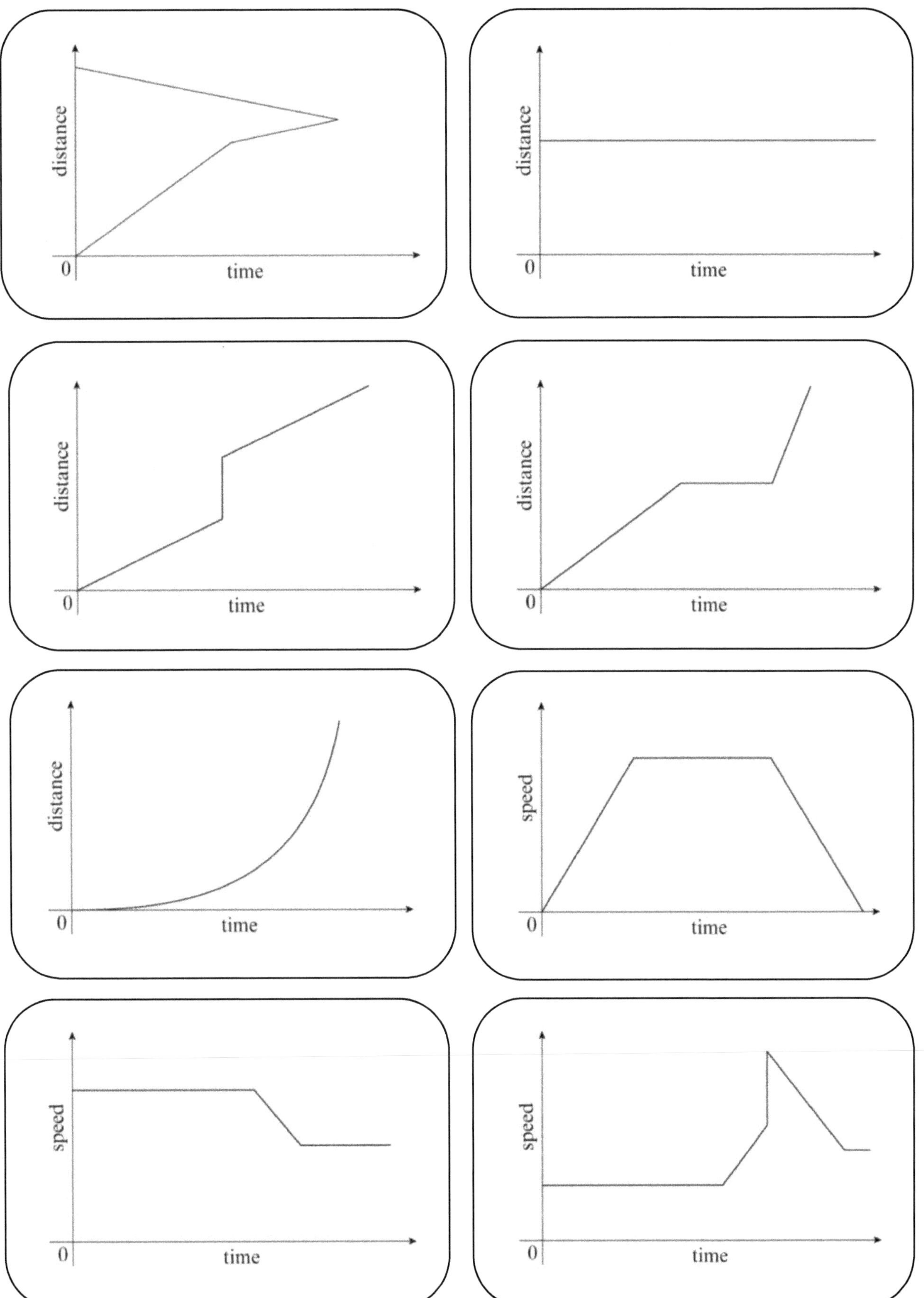

Rearranging formulae jigsaw

Teacher notes

This jigsaw activity is an interesting group work alternative to a written exercise. The advantages of a jigsaw activity over a written exercise are that it creates opportunities for students to discuss their ideas, and they can find an alternative starting point if they are not sure how to do a particular question. Students are able to start with the answer and work backwards, which is sometimes an interesting way to tackle a question as students could rearrange either expression.

This activity concentrates on rearranging formulae where the unknown, x, appears more than once. Students should already be familiar with these questions and in particular know in what order they should do operations – this is tested in this jigsaw.

Introductory activity

If you want to use this task as a revision activity, or a plenary, you may find that little introduction is necessary. Otherwise you may want to go through the following examples.

Example 1

Make x the subject of $ax = c - bx$

$ax = c - bx$	collect x's on one side
$ax + bx = c$	factorise LHS
$x(a+b) = c$	divide both sides by $(a + b)$
$x = \dfrac{c}{a+b}$	

Example 2

Make x the subject of $b = \dfrac{ax+c}{dx-3}$

$b = \dfrac{ax+c}{dx-3}$	multiply both sides by $(dx - 3)$
$b(dx-3) = ax + c$	multiply out brackets
$bdx - 3b = ax + c$	collect x's on one side and all terms without x on the other side
$bdx - ax = c + 3b$	factorise LHS
$x(bd - a) = c + 3b$	divide both sides by $(bd - a)$
$x = \dfrac{c+3b}{bd-a}$	

Example 3

Make x the subject of $a = \sqrt{\dfrac{b-x^2}{x^2}}$

$a = \sqrt{\dfrac{b-x^2}{x^2}}$	square both sides
$a^2 = \dfrac{b-x^2}{x^2}$	multiply by x^2
$a^2x^2 = b - x^2$	collect x's on one side
$a^2x^2 + x^2 = b$	factorise LHS
$x^2(a^2+1) = b$	divide both sides by $(a^2 + 1)$
$x^2 = \dfrac{b}{a^2+1}$	square root both sides
$x = \pm\sqrt{\dfrac{b}{a^2+1}}$	

Example 3

Make x the subject of $3 = \dfrac{a}{x} + b$

$3 = \dfrac{a}{x} + b$	subtract b from both sides
$3 - b = \dfrac{a}{x}$	multiply both sides by x
$x(3-b) = a$	divide both sides by $(3 - b)$
$x = \dfrac{a}{3-b}$	

Note: students also need to be aware that $x = \dfrac{2+b}{c+3}$ is the same as $x = \dfrac{b+2}{3+c}$, however $x = \dfrac{2-b}{c}$ is not the same as $x = \dfrac{b-2}{c}$.

Students need to be aware of the difference the minus sign makes. You may want to discuss this further.

Jigsaw activity

Ask your students to work in small groups. Copy the jigsaw pieces onto card to make a complete set of pieces for each group. There are several different-shaped jigsaws in the teacher resources so you should not tell the students what shape to expect. This jigsaw has blank edges around the outside which makes it a little easier for your students.

Differentiation

To make it more difficult you could write harder questions and answers around the blank outside edges of the jigsaw.

Alternative approaches

You could just ask students to complete the jigsaw or you could copy blank jigsaw pieces from the online resources and ask students to write their own "outside edges" to extend the jigsaw themselves – if you decide to do this you may want to copy the jigsaw pieces onto paper rather than card. Students could make their own new jigsaw using the blank pieces from the online resources.

Answers

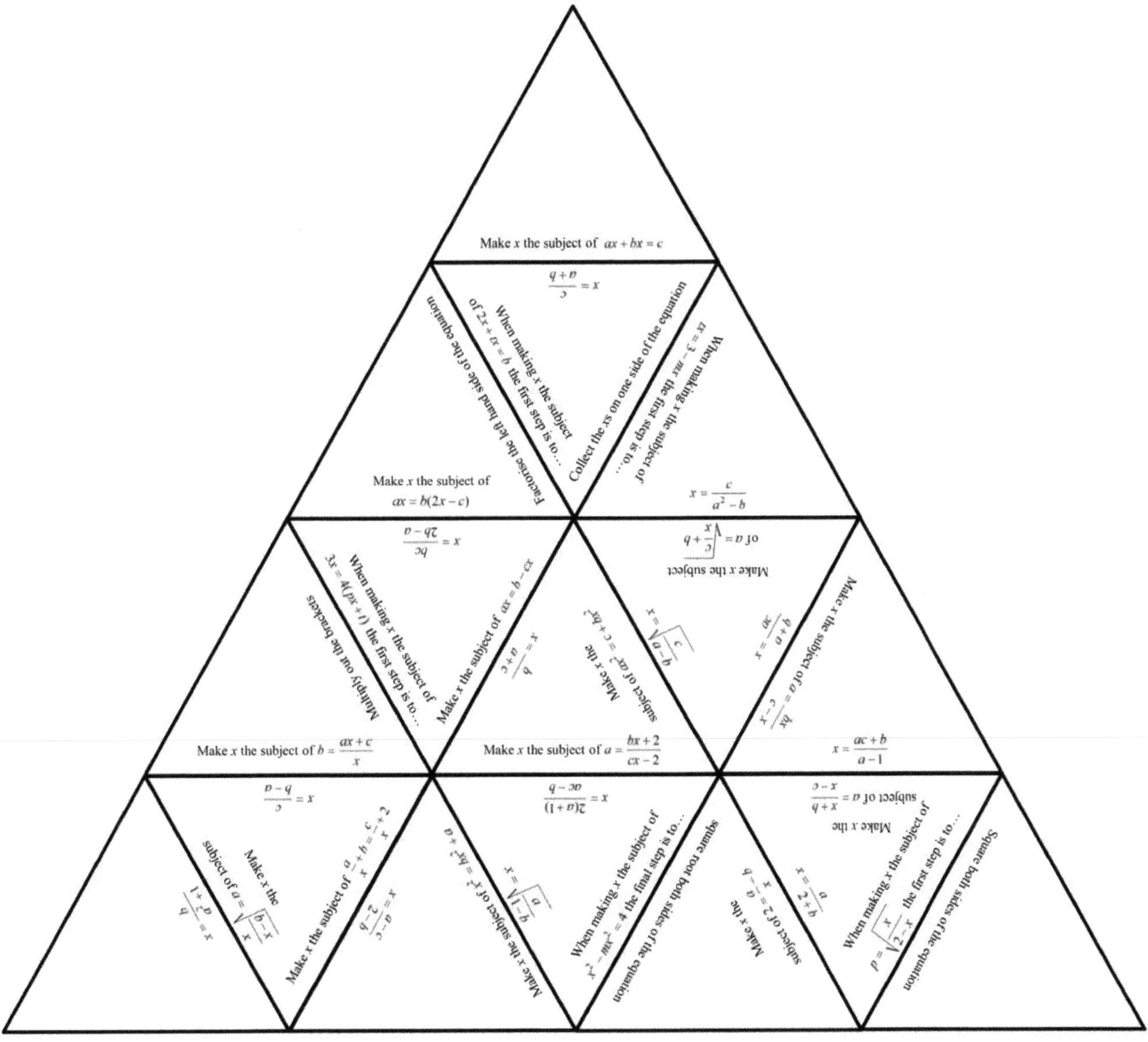

Rearranging formulae jigsaw pieces

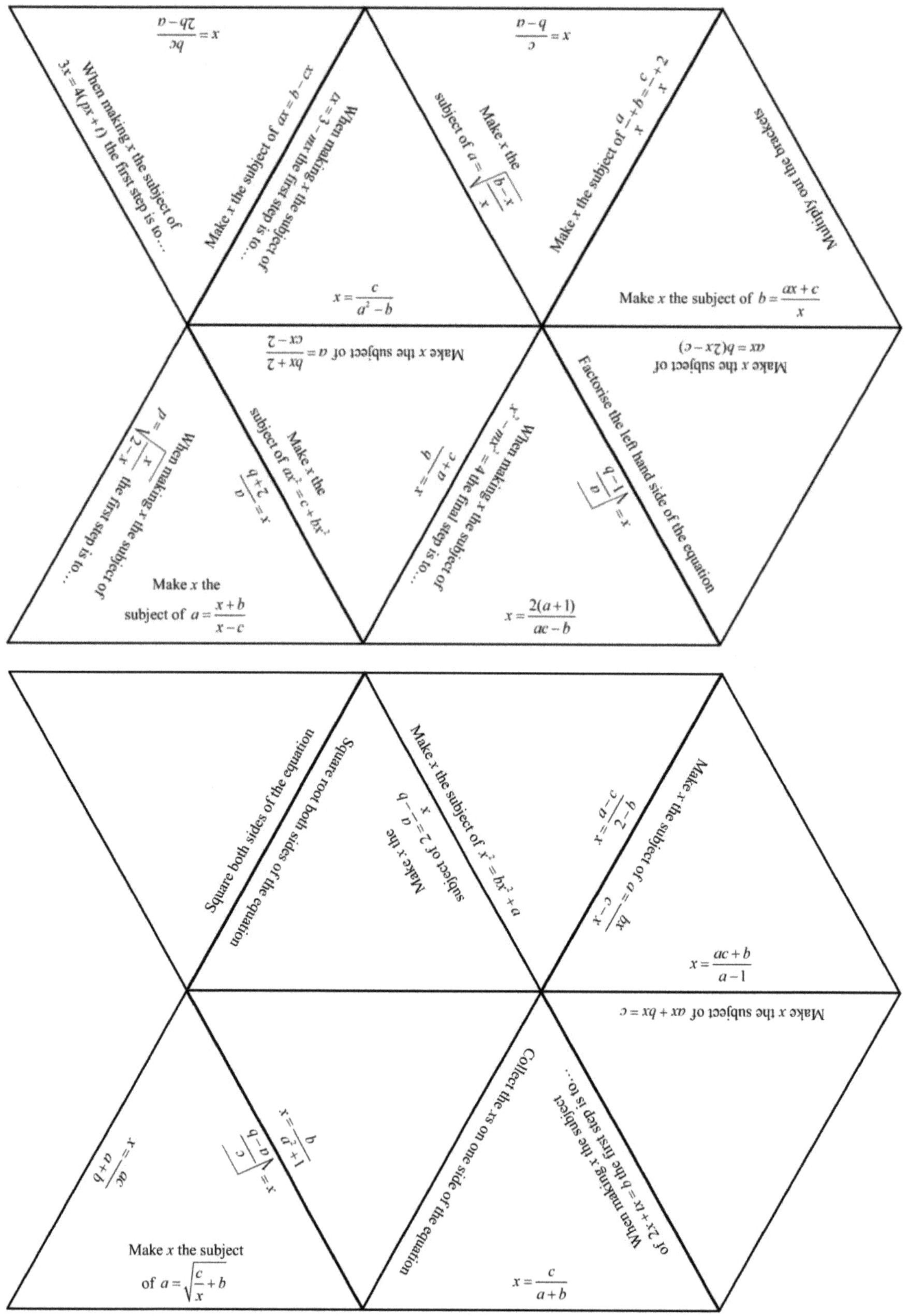

Larger size jigsaw pieces are available in the online resources.

Geometry crossword

Teacher notes

This activity is for students who already know most of their geometry facts and focuses on some of the key vocabulary from the syllabus. As a revision activity it covers many different areas of the syllabus including: geometrical terms and relationships, geometrical constructions, symmetry and angle properties.

Introductory activity

There should be little need for any introduction, as this is intended as a revision exercise.

Crossword activity

Print out one copy of the crossword grid and clues for your students, or pairs of students, and ask them to complete the crossword by filling in one letter per box.

Differentiation

Ask students to race to complete the crossword or ask them if they can devise alternative clues for some of the words.

Alternative approaches

Ask students to produce their own crossword. A blank crossword grid is available in the online resources.

Answers

1 P	E	2 R	P	E	N	D	I	3 C	U	L	A	R								
E		E						Y												
N		F						C			4 C				5 S					
T		L						L			6 E	X	T	E	R	I	O	R		
A		E		7 O				I			N				M					
G		X		8 B	I	S	E	C	T		T				I					
O				T							R		9 C		L			10 F		
N				U			11 P				E		12 O	C	T	A	G	O	N	
				S			A						R		R			U		
13 A	C	U	14 T	E			15 R	E	G	U	L	A	R					R		
			W				A						E						16 A	
	17 I	S	O	S	C	E	L	E	S				S		18 E				L	
	R						L						P		Q				T	
	R				19 T		20 E	Q	U	A	L		O		U		21 N		E	
	E				A		L						N		I		I		R	
	G		22 H		N								D		L		N		N	
	U		E		G			23 Q	U	A	D	R	I	L	A	T	E	R	A	L
	L		X		E								N		T		T		T	
	A		A		N								G		E		Y		E	
	24 R	I	G	H	T	A	N	G	L	E					R					
			O									25 T	R	I	A	N	G	L	E	
	26 C	O	N	G	R	U	E	N	T						L					

Geometry crossword

Geometry crossword clues

Across

1 Two straight lines at right angles to each other are .....

6 The ..... angles of all regular polygons add up to 360°.

8 To cut an angle exactly in half.

12 An eight sided shape.

13 An angle smaller than 90°.

15 Polygons where all the sides are equal and all interior angles are equal.

17 A triangle with two equal sides.

20 In a circle the angles in the same segment are .....

23 A four sided shape.

24 90°.

25 A three sided shape.

26 When two shapes are exactly the same shape and size they are .....

Down

1 The exterior angle of a regular .... is 72°.

2 An angle bigger than 180°.

3 A quadrilateral that has all four corners on the circumference of a circle is a ..... quadrilateral.

4 The perpendicular bisector of a chord passes through the ..... of the circle.

5 When one shape is an enlargement of another the shapes are .....

7 An angle between 90° and 180°.

9 Equal angles formed by two parallel lines and a crossing transversal line, the equal angles are on the same side of the transversal (sometimes known as F angles).

10 The order of rotational symmetry of a square.

11 Two straight lines that will never cross each other.

14 The number of lines of symmetry of a rectangle.

16 Equal angles formed by two parallel lines and a crossing transversal line, the equal angles are on opposite sides of the transversal (sometimes known as Z angles).

17 Polygons where all the sides are not equal in length.

18 A triangle with all three angles of 60°.

19 A straight line which just touches a curve at one given point.

21 The angle on the circumference of a circle in a triangle based on the diameter is ..... degrees.

22 A six sided shape.

Cumulative frequency – three in a row game

Teacher notes

Cumulative frequency is a topic students can sometimes find a little “dry”. The purpose of this activity is to make the topic more enjoyable and engaging and also consolidate students’ basic knowledge. The activity covers the shape of the cumulative frequency curve, median, quartiles, percentiles, inter-quartile range, working out cumulative frequencies and reading values from cumulative frequency graphs.

Introductory activity

This activity is intended as a revision of the basics and is probably best suited to later on in the course. Or you could also use it as a fun alternative to a student book exercise in which case you may need to go through the following examples.

Example 1

To find the cumulative frequencies, calculate the running total of the frequencies.

x	f
$0 < x \leq 10$	4
$10 < x \leq 20$	7
$20 < x \leq 30$	10
$30 < x \leq 40$	4
$40 < x \leq 50$	1

x	cf
$0 \leq x \leq 10$	4
$0 \leq x \leq 20$	11
$0 \leq x \leq 30$	21
$0 \leq x \leq 40$	25
$0 \leq x \leq 50$	26

Example 2

To draw the curve, start by plotting the lower limit of the first class against 0. Then plot all upper limits against the cumulative frequencies. Join the points with a smooth curve, this will be S-shaped.

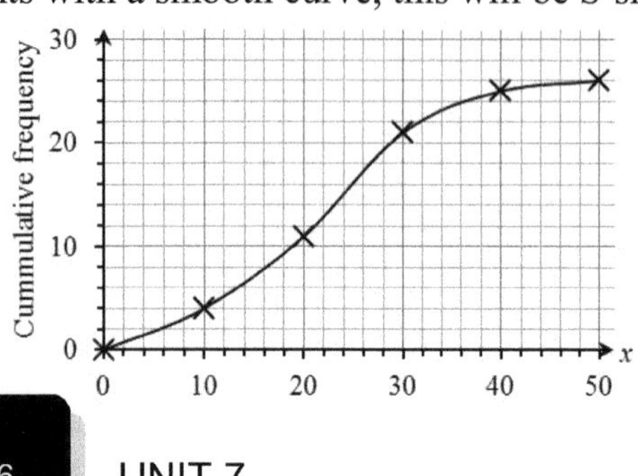

Example 3

To Find the median and quartiles and percentiles proceed as follows.

There are 26 values in total, halve this to find the position of the median (13) and read off the x-value from the graph.

Find $\frac{1}{4}$ of 26 (6.5) for the position of the lower quartile and $\frac{3}{4}$ of 26 (19.5) for the position of the upper quartile.

Find 90% of 26 (23.4) for the position of the 90th percentile.

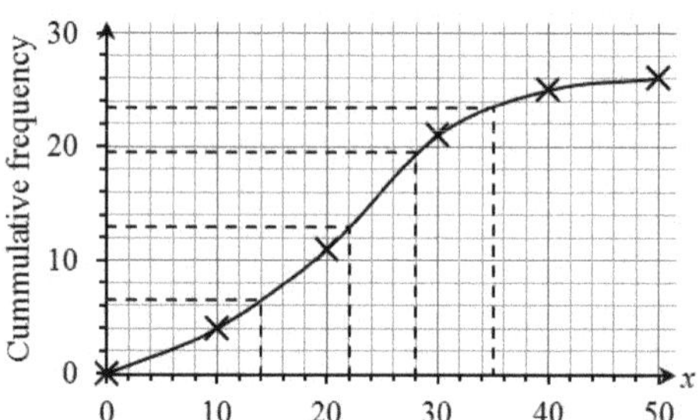

In this example the estimates are:

Median = 22 LQ = 14 UQ = 28
90th percentile = 35
(IQR = 28 – 14 = 14)

You could give an example on reading off a graph from the horizontal axis. For example, find how many people had a time of less than 20 seconds.

Three in a row activity

For this activity you need two players, or two teams, each with a different set of coloured counters. For each team you will need one copy of the game board, which is in two halves left hand side and right hand side, and a set of 24 cards.

Shuffle the cards and deal 12 to each player. The youngest player turns over their first card and matches their card to a space on the game board. They place their coloured counter on that space and the card on a discard pile. Then it is the next player's turn to do the same. If a player makes an incorrect matching then their counter is removed and their opponent gets a bonus go with that card before completing their normal turn. This is a reward for being vigilant and spotting mistakes.

The winner is the first player to get three coloured counters in a row horizontally, vertically or diagonally.

Differentiation

You could ask for students to get two counters or four counters in a row to make it easier or harder (which will also make the game shorter or longer).

Alternative approaches

You could introduce more players and ask students to come up with their own rules to play the game. You could ask students to produce their own cumulative frequency board game using the blank copy of the board and game cards included in the online resources.

Answers

The game cards are printed in the same order as the game board (working from top left to bottom right in both cases). This will not be an issue when they play the game, provided that the cards are shuffled.

Cumulative frequency game cards set 1

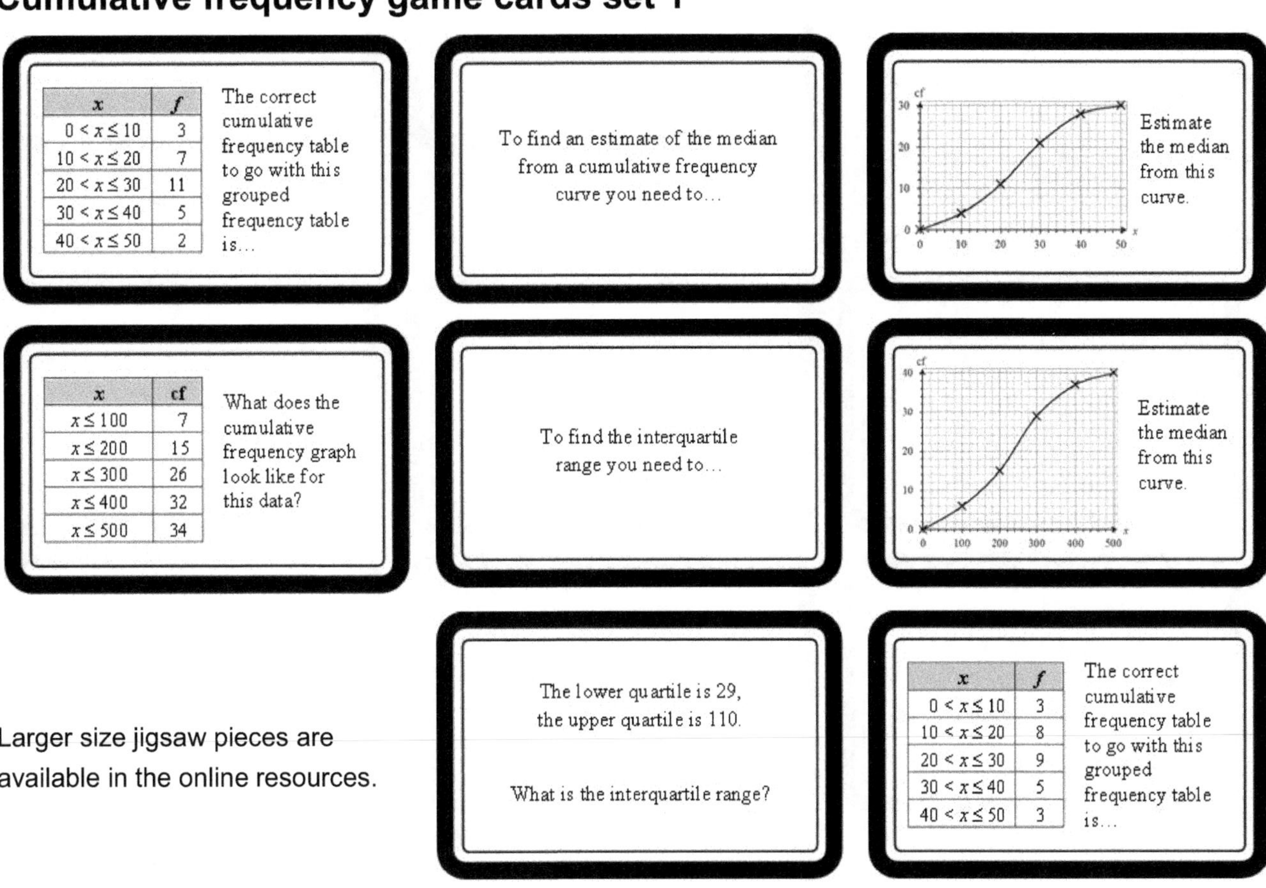

x	f
$0 < x \leq 10$	3
$10 < x \leq 20$	7
$20 < x \leq 30$	11
$30 < x \leq 40$	5
$40 < x \leq 50$	2

The correct cumulative frequency table to go with this grouped frequency table is...

To find an estimate of the median from a cumulative frequency curve you need to...

Estimate the median from this curve.

x	cf
$x \leq 100$	7
$x \leq 200$	15
$x \leq 300$	26
$x \leq 400$	32
$x \leq 500$	34

What does the cumulative frequency graph look like for this data?

To find the interquartile range you need to...

Estimate the median from this curve.

The lower quartile is 29, the upper quartile is 110.

What is the interquartile range?

x	f
$0 < x \leq 10$	3
$10 < x \leq 20$	8
$20 < x \leq 30$	9
$30 < x \leq 40$	5
$40 < x \leq 50$	3

The correct cumulative frequency table to go with this grouped frequency table is...

Larger size jigsaw pieces are available in the online resources.

Cumulative frequency game cards set 2

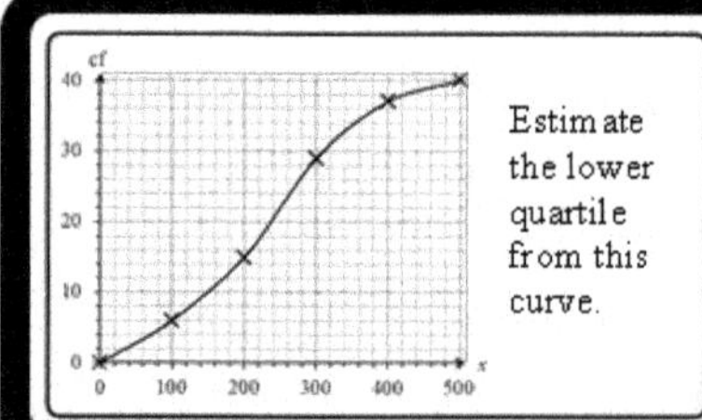

Estimate the lower quartile from this curve.

To find an estimate of the lower quartile from a cumulative frequency curve you need to…

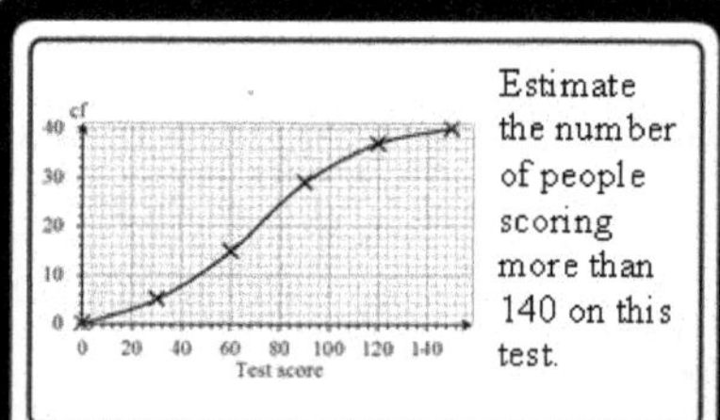

Estimate the number of people scoring more than 140 on this test.

Percentiles divide the cumulative frequencies into…

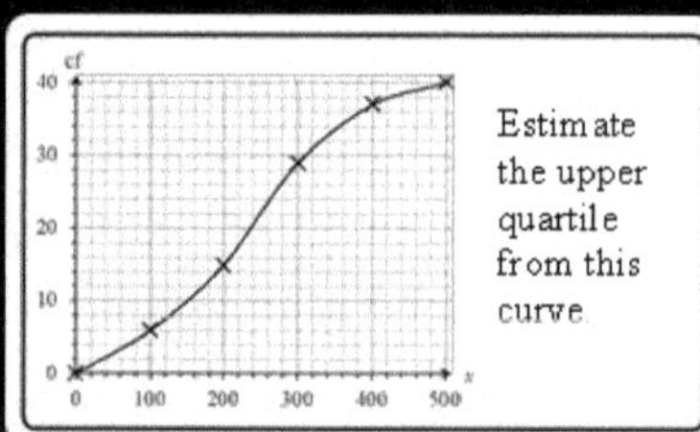

Estimate the upper quartile from this curve.

x	cf
$x \le 50$	70
$x \le 100$	150
$x \le 150$	260
$x \le 200$	320
$x \le 250$	340

What does the cumulative frequency graph look like for this data?

Larger size jigsaw pieces are available in the online resources.

To find an estimate of the upper quartile from a cumulative frequency curve you need to …

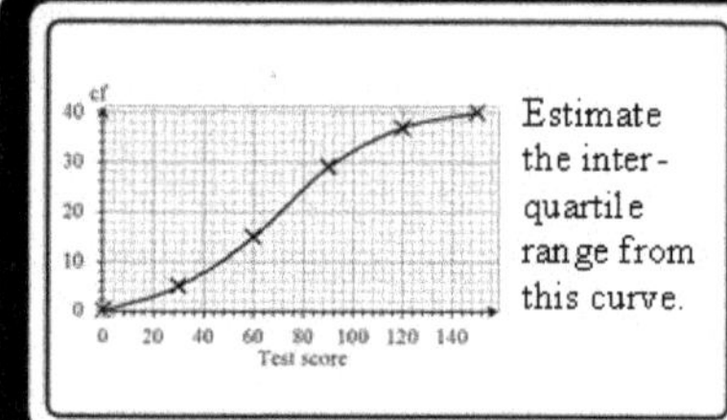

Estimate the inter-quartile range from this curve.

Cumulative frequency game cards set 3

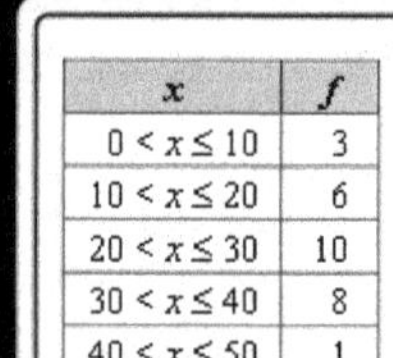

x	f
$0 < x \le 10$	3
$10 < x \le 20$	6
$20 < x \le 30$	10
$30 < x \le 40$	8
$40 < x \le 50$	1

The correct cumulative frequency table to go with this grouped frequency table is…

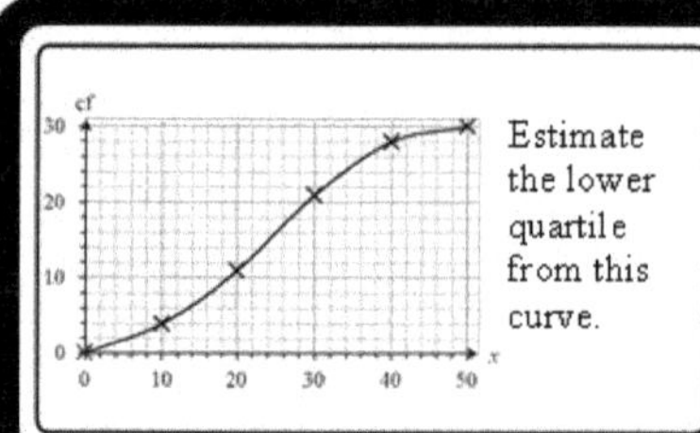

Estimate the lower quartile from this curve.

Estimate the number of people scoring less than 140 on this test.

x	cf
$x \le 50$	7
$x \le 100$	15
$x \le 150$	26
$x \le 200$	32
$x \le 250$	34

What does the cumulative frequency graph look like for this data?

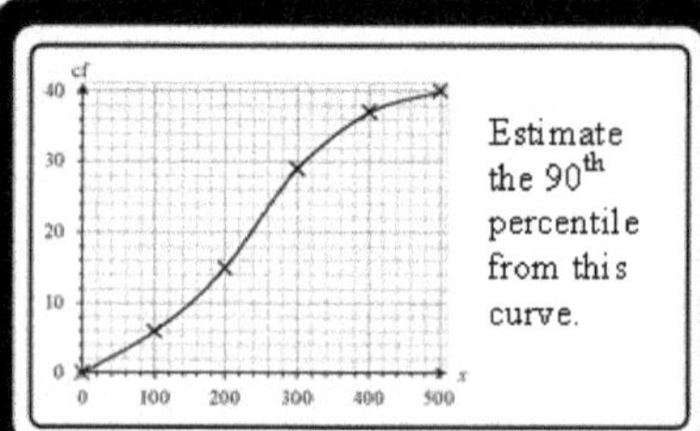

Estimate the 90th percentile from this curve.

The lower quartile is 17, the upper quartile is 96.

What is the interquartile range?

Larger size game cards are available in the online resources.

Quartiles divide the cumulative frequencies into …

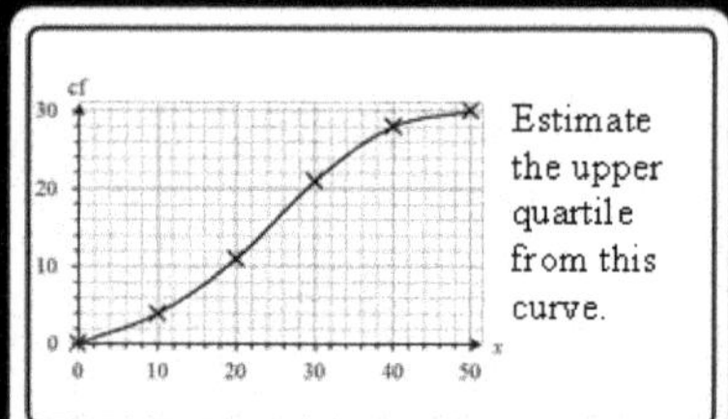

Estimate the upper quartile from this curve.

Cumulative frequency right hand side game board

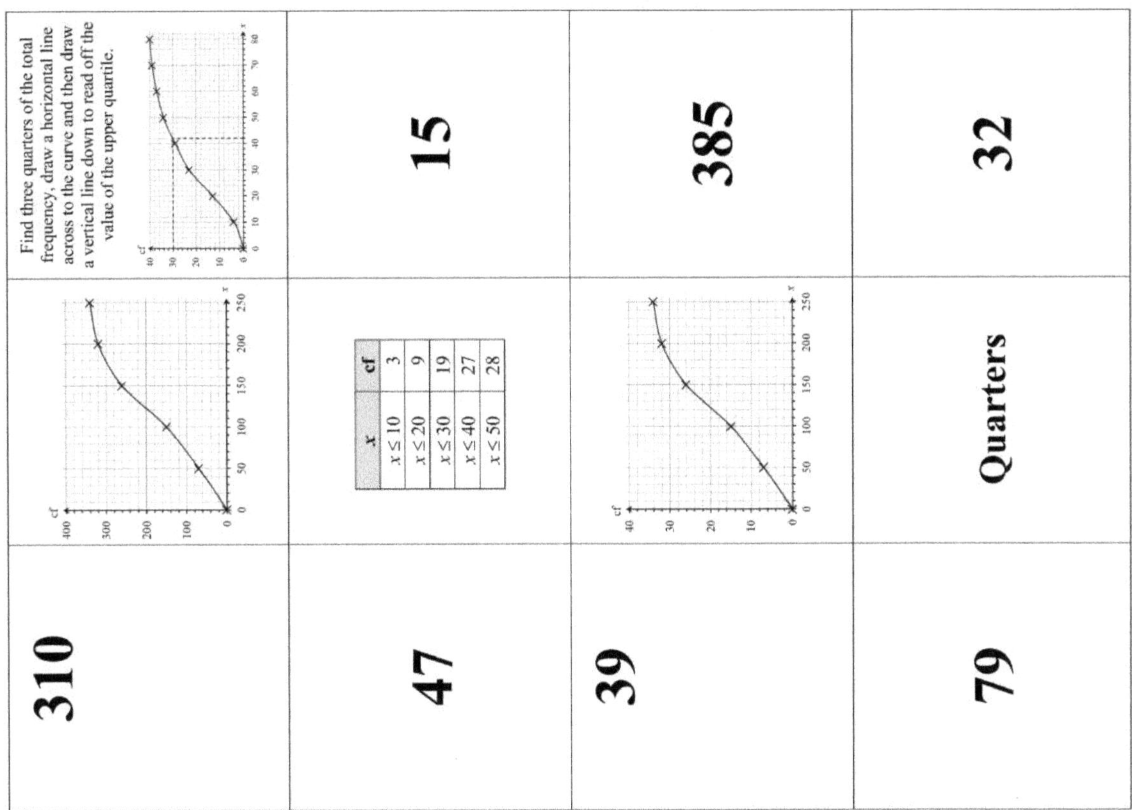

Find three quarters of the total frequency, draw a horizontal line across to the curve and then draw a vertical line down to read off the value of the upper quartile.
15
385
32
x
cf
x ≤ 10
3
x ≤ 20
9
x ≤ 30
19
x ≤ 40
27
x ≤ 50
28
Quarters
310
47
39
79

Cumulative frequency left hand side game board

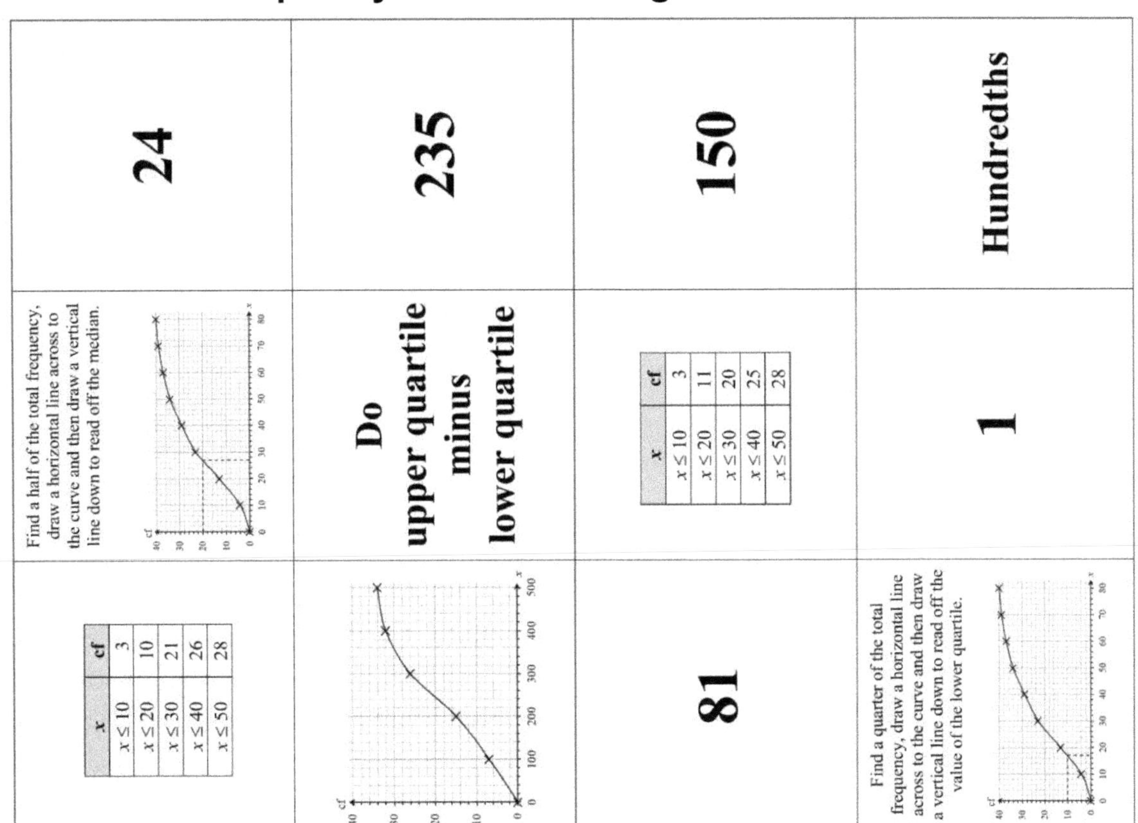

24
235
150
Hundredths
Find a half of the total frequency, draw a horizontal line across to the curve and then draw a vertical line down to read off the median.
Do
upper quartile
minus
lower quartile
x
cf
x ≤ 10
3
x ≤ 20
11
x ≤ 30
20
x ≤ 40
25
x ≤ 50
28
1
x
cf
x ≤ 10
3
x ≤ 20
10
x ≤ 30
21
x ≤ 40
26
x ≤ 50
28
81
Find a quarter of the total frequency, draw a horizontal line across to the curve and then draw a vertical line down to read off the value of the lower quartile.

Larger size jigsaw pieces are available in the online resources.

Algebraic fractions jigsaw

Teacher notes

This jigsaw activity is an interesting group work alternative to a written exercise. The advantages of a jigsaw activity over a written exercise are that it creates opportunities for students to discuss their ideas, and they can find an alternative starting point if they are not sure how to do a particular question. Students are able to start with the answer and work backwards, which is sometimes an interesting way to tackle a question.

This activity requires students to be able to simplify algebraic expressions by using either basic division, the laws of indices or by factorising. Students need to be able to factorise quadratics into either one bracket, for example, $x^2 + 3x = x(x + 3)$, or two brackets, for example, $x^2 - 9x + 20 = (x - 5)(x - 4)$, including the difference of two squares, for example $x^2 - 4 = (x - 2)(x + 2)$, and they need to be able to cancel algebraic fractions.

Introductory activity

If you want to use this as a revision activity you may find that little more introduction is necessary. Otherwise you may want to go through the following examples.

Example 1

$$\frac{8}{4x} = \frac{2}{x}$$

Often students wrongly cancel this as $\frac{8x}{4} = 2x$

Example 2

$$\frac{8x^2 \times 2x^3}{10x^4} = \frac{16x^5}{10x^4} = \frac{8x}{5}$$

Using the laws of indices $x^a \times x^b = x^{a+b}$ and $x^a \div x^b = x^{a-b}$

Example 3

$$\frac{x^2 - 9}{x + 3} = \frac{(x-3)(x+3)}{x+3} = x - 3$$

Use the difference of two squares to factorise the numerator.

Example 4

$$\frac{x^2 + 7x + 12}{x^2 + x - 6} = \frac{(x+3)(x+4)}{(x+3)(x-2)} = \frac{x+4}{x-2}$$

Remind students that when cancelling fractions such as these they need to factorise the numerator and denominator and then look for the common factor, in this case $(x + 3)$ to cancel out.

Jigsaw activity

Ask your students to work in small groups. Copy the jigsaw pieces onto card to make a complete set for each group. There are several different-shaped jigsaws in the Teacher Resource Kit so you should not tell the students what shape to expect. This jigsaw does not have any blank edges which mean that it is harder for your students to work out the final shape.

Differentiation

To make it more difficult there are questions or 'solutions' around the outside edges of the jigsaw. To make it easier, you could indicate outside edges; perhaps draw a bold line along them.

Alternative approaches

You could just ask students to complete the jigsaw or you could make use of the fact the outside edges are not blank. You could copy blank jigsaw pieces from the online resources and ask students to use these pieces to extend the jigsaw, either finding the solutions to the harder questions, or writing a question that has the given answer, or students could make an entirely new jigsaw of their own.

Answers

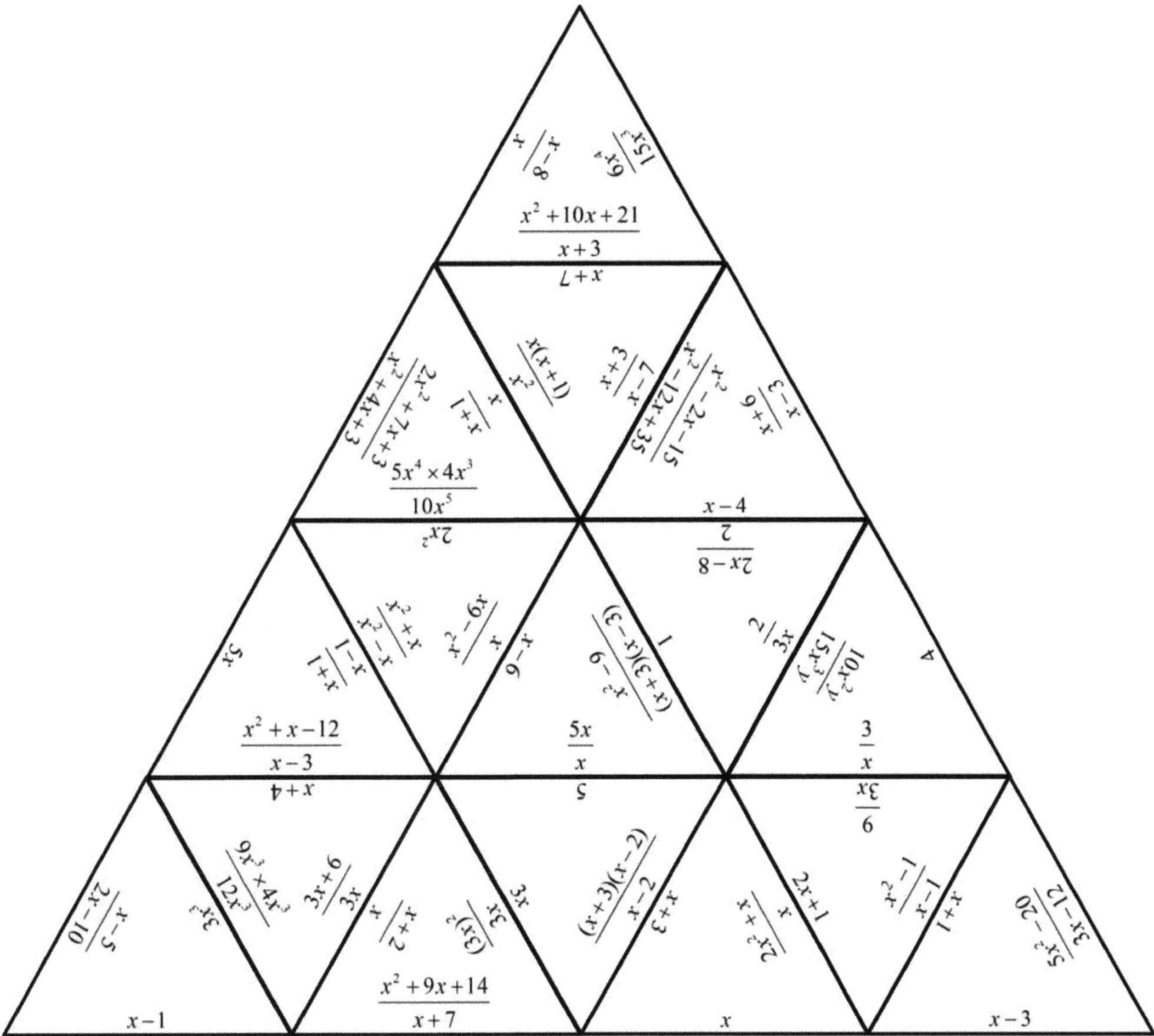

Algebraic fractions jigsaw pieces

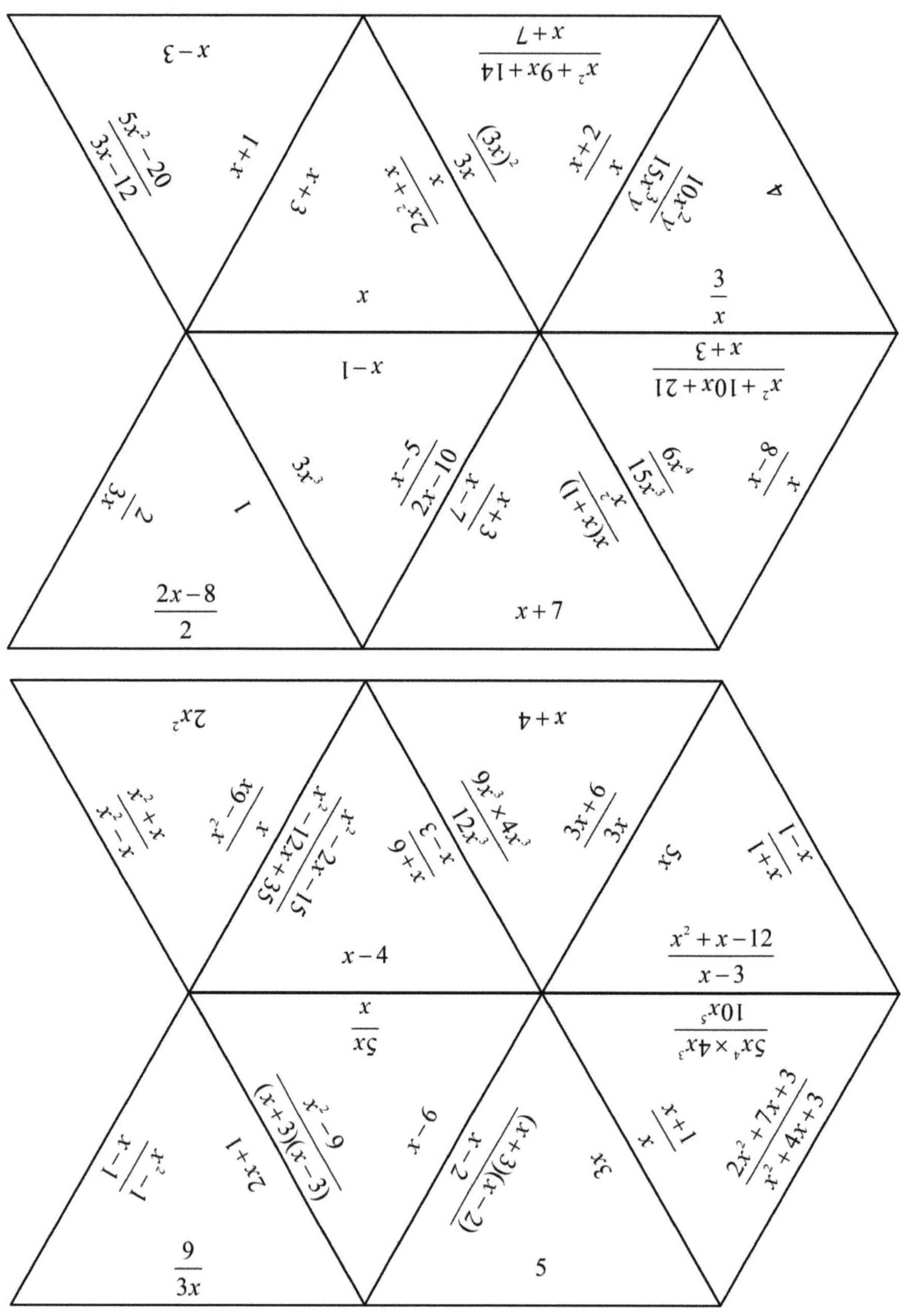

Larger size jigsaw pieces are available in the online resources.

Teacher notes

IGCSE students often perform poorly in direct and inverse variation problems. Students miss valuable method marks by poor layout of working, poor methods or a general lack of understanding. The purpose of this activity is to help students understand the best way to layout working for direct and inverse variation questions, and to ensure that all stages of the calculation are included to get both method and accuracy marks. Students get confused between direct and inverse variation and the examples included will help distinguish the difference.

Introductory activity

You will need to discuss the steps needed to solve direct and inverse variation questions.

Step 1

Start with the proportion symbol and then turn the statement into an equation: "p varies directly as q" becomes $p \propto q$ which then becomes $p = kq$.

Step 2

Substitute the values of p and q given in the question into the equation to find the value of k.

Step 3

Use this value of k in the equation to find the missing value.

All questions follow the same three basic steps, but students must realise the difference between direct variation $p \propto q$ and inverse variation $p \propto 1/q$ and how to deal with more complex relationships.

For example,

p varies directly as the square root of $(q - 5)$ is

$$p \propto \sqrt{(q-5)}$$

p varies inversely as the square $(q - 5)$ is

$$p \propto \frac{1}{(q-5)^2}$$

Ordering activity

Ask your students to work in small groups. Photocopy the four calculations onto card and cut them out, to make one complete set for each group of students. Ask students to sort them into four correct calculations with correct working and order of working shown for each.

Differentiation

To make this task easier ask the students to work through just one calculation group of eight cards first, putting them in order, starting with the question and ending with the answer. To make the task harder use all four groups of eight mixed together but miss out some cards. Students will need to identify the missing cards.

Alternative approaches

You could use the blank cards provided in the online resources to ask students to write their own set of cards to correctly order the working for their own chosen problem. Encourage using eight cards, one for the question and seven for each step of working.

Answers

Calculation 1

The quantity p varies inversely as the square root of $q + 2$. $p = 5$ when $q = 7$. Find p when $q = 14$.

1 $p \propto \frac{1}{\sqrt{(q+2)}}$ **2** $p = \frac{k}{\sqrt{(q+2)}}$

3 $5 = \frac{k}{\sqrt{(7+2)}}$ **4** $k = 15$

5 $p = \frac{15}{\sqrt{(q+2)}}$ **6** $p = \frac{15}{\sqrt{(14+2)}}$

7 $p = 3\frac{3}{4}$

Calculation 2

The quantity p varies directly as the quantity q. $p = 5$ when $q = 3$. Find p when $q = 8$.

1 $p \propto q$ **2** $p = kq$

3 $5 = 3k$ **4** $k = \frac{5}{3}$

5 $p = \frac{5}{3}q$ **6** $p = \frac{5}{3} \times 8$

7 $p = 13\frac{1}{3}$

Calculation 3

The quantity p varies directly as the square of $q + 2$. $p = 5$ when $q = 3$. Find p when $q = 8$.

1 $p \propto (q+2)^2$ 2 $p = k(q+2)^2$

3 $5 = k(3+2)^2$ 4 $k = 0.2$

5 $p = 0.2(q+2)^2$ 6 $p = 0.2(8+2)^2$

7 $p = 20$

Calculation 4

The quantity p varies inversely as the square of $q + 2$. $p = 5$ when q = 3. Find p when $q = 8$.

1 $p \propto \frac{1}{(q+2)^2}$ **2** $p = \frac{k}{(q+2)^2}$

3 $5 = \frac{k}{(3+2)^2}$ **4** $k = 125$

5 $p = \frac{125}{(q+2)^2}$ **6** $p = \frac{125}{(8+2)^2}$

7 $p = 1.25$

Direct and inverse variation sort cards

Calculation 1

$p = \frac{k}{\sqrt{(q+2)}}$	$k = 15$
$p = 3\frac{3}{4}$	$p = \frac{15}{\sqrt{(q+2)}}$
The quantity p varies inversely as the square root of $q + 2$. $p = 5$ when $q = 7$. Find p when $q = 14$.	$p = \frac{15}{\sqrt{(14+2)}}$
$5 = \frac{k}{\sqrt{(7+2)}}$	$p \propto \frac{1}{\sqrt{(q+2)}}$

Calculation 2

The quantity p varies directly as the quantity q. $p = 5$ when $q = 3$. Find p when $q = 8$.	$k = \frac{5}{3}$
$p = \frac{5}{3} \times 8$	$p \propto q$
$p = kq$	$p = 13\frac{1}{3}$
$p = \frac{5}{8}q$	$5 = 3k$

Calculation 3

$p = 20$	$p = 0.2(8+2)^2$
$5 = k(3+2)^2$	The quantity p varies directly as the square of $q + 2$. $p = 5$ when $q = 3$. Find p when $q = 8$.
$p \propto (q+2)^2$	$p = k(q+2)^2$
$k = 0.2$	$p = 0.2(q+2)^2$

Calculation 4

$p = \frac{125}{(q+2)^2}$	$5 = \frac{k}{(3+2)^2}$
$p = 1.25$	The quantity p varies inversely as the square of $q + 2$. $p = 5$ when $q = 3$. Find p when $q = 8$.
$p = \frac{k}{(q+2)^2}$	$p = \frac{125}{(8+2)^2}$
$k = 125$	$p \propto \frac{1}{(q+2)^2}$

Sine and cosine rule and area of a triangle correct my homework activity

Teacher notes

We often ask our students to make sure that they have checked their working, but this is a task that we rarely teach them how to do. The purpose of this activity is to help students understand what they should focus on when checking their working. It is always a good idea to write down every figure on the calculator display before rounding, then if a rounding error is made in an exam, the student will still gain all accuracy marks. Students should also realise there can be more than one way to calculate an angle and so they can check their working by using an entirely different method for calculating the missing angles. To understand this task, students need to be able to use both the cosine and sine rules to find a missing angle and remember the basic fact that all angles in a triangle add up to 180°. They also need to know how to use the formula for working out the area of a non-right angle triangle.

Introductory activity

Too much introduction and explanation will make this task too easy for the students. The main errors in the working given are related to premature/ incorrect rounding or using incorrect side lengths in the formulae. You may want to do this as an activity immediately following on from the work on sine rule and cosine rules in the student book or as a revision activity later in the course. You could remind students how to label a triangle (capital letters for the vertices and lower case letters for the edges opposite) and remind them of the sine and cosine rules and area of a triangle rule.

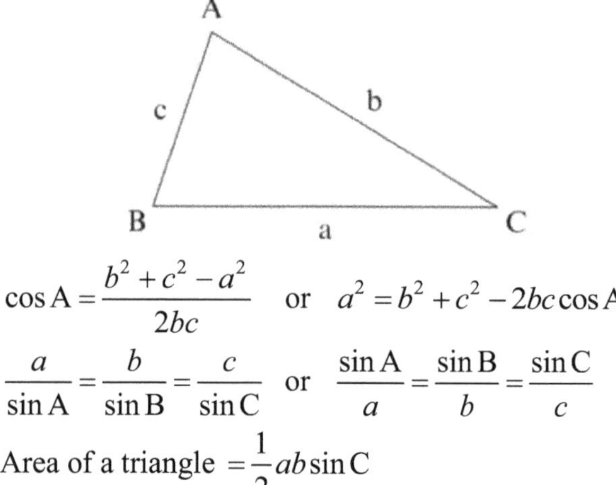

$\cos A = \dfrac{b^2 + c^2 - a^2}{2bc}$ or $a^2 = b^2 + c^2 - 2bc\cos A$

$\dfrac{a}{\sin A} = \dfrac{b}{\sin B} = \dfrac{c}{\sin C}$ or $\dfrac{\sin A}{a} = \dfrac{\sin B}{b} = \dfrac{\sin C}{c}$

Area of a triangle $= \dfrac{1}{2}ab\sin C$

Correct my homework activity

Ask your students to work in small groups or pairs for this activity so that they can discuss ideas. Give each group or pair a copy of the "Correct my homework activity" sheet. Ask students to use the right-hand column in the table to write down an explanation for the working that has been done and to make any necessary corrections.

Differentiation

To make this task a little easier, give the students a few of the answers in the second column. To make this task a bit harder, miss out some of the lines of working in the first column. For example, miss out the lines

$\cos A = \dfrac{85}{140}$ and $\sin B = \dfrac{7 \times \sin 52}{8}$

Alternative approaches

Ask students to solve their own question on sine and/or cosine rules, making deliberate mistakes. They can then give this to a different group to explain and correct.

Answers

Q	Explanation/Corrections
a)	The student has drawn a correct sketch of the triangle. The student is indicating that they are trying to find angle A, however this is the first mistake as angle B is the largest angle. The largest angle is opposite the longest side.
	The cosine rule is correctly used to find angle A, even though this isn't the biggest angle.
	The student has simplified the numerator and denominator correctly.
	The student is correctly using inverse cosine to find the angle.
	The correct angle A is 52.6°, angles should be rounded to 1 d.p. or 3 s.f. The student has truncated the number. It is good practice to write down all figures from a calculator display $A = 52.61680158°$ before rounding.
b)	The student has correctly re-drawn the triangle with the angle marked (although this is still the wrong angle). The student has indicated they are now looking for angle B.
	The sine rule is incorrectly used to calculate angle B it should read. $\frac{\sin B}{10} = \frac{\sin 52.6}{8}$ so $B = \sin^{-1}\left(\frac{10 \times \sin 52.6}{8}\right) = 83.3°$ $B = 83.33457274°$ which is the largest angle. Note 52.61680158° should be used in this calculation not 52.6°!
	This line follows correctly from the line above: it is a correct rearrangement of their figures.
	This has been prematurely rounded to an inaccurate figure. At this point all the figures on the calculator display should be used.
	Finding the inverse sine is correctly done from the premature rounding.
	This follows correctly from the previous working but has been over rounded again.
	The student is using angles in a triangle add up to 180°, clearly if angle A was the largest angle then angle C would not be 84°. The correct working is: $C = 180 - (52.6168 + 83.3346) = 44.0486° = 44.1°$ Note if 3 s.f. answers used rather than exact answers you would get the wrong third angle for example, $C = 180 - (52.6 + 83.3) = 44.1°$ $C = \sin^{-1}\left(\frac{7 \times \sin 52.6}{8}\right) = 44.0°$ Check: $\frac{\sin C}{7} = \frac{\sin 52.6}{8}$
c)	The student has indicated that they intend to use angle A to work out the area of the triangle, so they should use the rule: Area of triangle $= \frac{1}{2}bc\sin A$
	There are two errors here: first the student has used 52° instead of the unrounded 52.61680158°; second the student has calculated , Area of triangle $= \frac{1}{2}ca\sin B$ The student should use the two side lengths that enclose the angle in the formula so the working should be: Area of triangle $= \frac{1}{2} \times 7 \times 10 \times \sin 52.61680158 = 27.81074433$
	This follows from their working, but the answer is wrong because of using the 8 instead of the 10 on the previous line. The student has not rounded their answer to 3 s.f. The correct answer is: Area of the triangle = 27.8 cm^2 (3 s.f.)

Sine and cosine rules and area of a non-right angle triangle correct my homework

In the working below there are some deliberate mistakes. Your task is to explain what is happening in the working that this student has shown and to correct all the errors. The first box has been started for you.

In ΔABC, $AB = 7$ cm, $AC = 10$ cm and $BC = 8$ cm.

a) Calculate the size of the largest angle.

b) Solve the triangle. (This means find all the missing sides and angle; here it is just the missing angles.)

c) Find the area of this triangle.

Working	Explanation/Corrections
a) A, B, C, 7, 10, 8	The student has drawn a correct sketch of the triangle. The student has indicated that they are trying to find angle A ....
$\cos A = \dfrac{7^2 + 10^2 - 8^2}{2 \times 7 \times 10}$	
$\cos A = \dfrac{85}{140}$	
$A = \cos^{-1}(85 \div 140)$	
The largest angle is $A = 52°$	
b) A, B, C, 52°, 7, 10, 8	
$\dfrac{\sin B}{7} = \dfrac{\sin 52}{8}$	
$\sin B = \dfrac{7 \times \sin 52}{8}$	
$\text{Sin } B = 0.69$	
$B = \sin^{-1}(0.69)$	
$B = 44°$	
$C = 180 - (52 + 44) = 84°$	
c) A, B, C, 52°, 7, 10, 8	
$\frac{1}{2} \times 7 \times 8 \times \sin 52$	
Area of triangle $= 22 \text{ cm}^2$	

A larger version is available in the online resources.

Histogram spot the mistake poster

Teacher notes

In assessments we often ask our students to make sure that they have checked their working. This is a task that we sometimes forget to teach them how to do. The purpose of this activity is to help students understand the sort of things they should concentrate on when checking their working in histogram questions. For example, it is a very common mistake for students to use heights of bars or frequencies instead of areas of bars and frequency densities, not realising the significance of the differing class widths. It is also very common for students to misread scales especially when one small square on an axis does not stand for 1 unit or 0.1 units. Both of these issues are addressed in the poster.

Introductory activity

This is best used as a revision activity and requires no introduction if you use it immediately after studying the histograms work in unit 8. You could remind students about the formula:

Frequency = frequency density × class width

Spot the mistake activity

Ask your students to work in small groups or pairs for this activity so that they can discuss ideas. Give each group (or pair) a copy of the spot the mistake poster. Ask students to annotate the poster where they find any mistakes.

Differentiation

To make it easier you could tell students how many mistakes the poster includes and the common mistakes that students make.

To extend the activity you could ask students to correct the mistakes and produce the correct solutions to the problems.

Alternative approaches

Ask students to make their own spot the mistake histogram poster. Students can then share this incorrect work with a different group for that group to spot the mistakes.

Histogram spot the mistake

In this poster on histograms there are some deliberate mistakes. Your task is to find as many as you can.

Question 1

The histogram shows the distribution of the masses of cases on a flight.

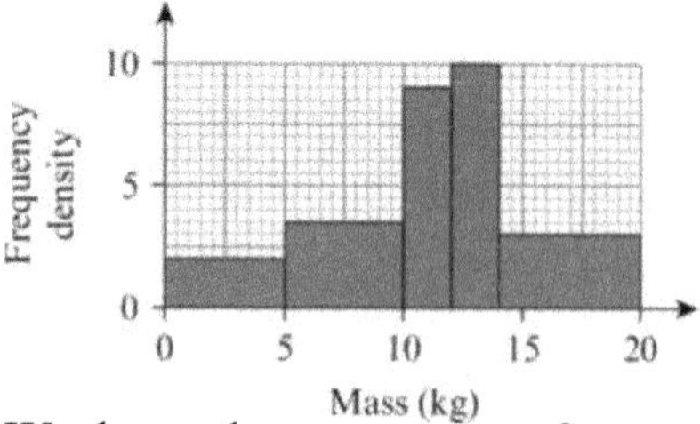

Work out the percentage of cases with a mass of greater than 12 kg.

Total number of cases $= 2 + 3 + 9 + 10 + 3 = 27$

Number of cases with a mass greater than 12 is $10 + 3 = 13$

Percentage of cases with a mass greater than 12 is

$\frac{13}{27} \times 100 = 48.1\%$ to 3 significant figures.

Question 2

The table gives information about the areas of pictures used in newspapers.

Area (A cm^2)	Frequency
$0 < A \leq 10$	38
$10 < A \leq 25$	36
$25 < A \leq 40$	30
$40 < A \leq 60$	46

Draw a histogram to represent this information.

Working:

Area (A cm^2)	Frequency	CW	FD
$0 < A \leq 10$	30	10	3
$10 < A \leq 25$	35	5	7
$25 < A \leq 40$	75	15	5
$40 < A \leq 60$	20	10	2

Histogram to show area of pictures used in newspapers:

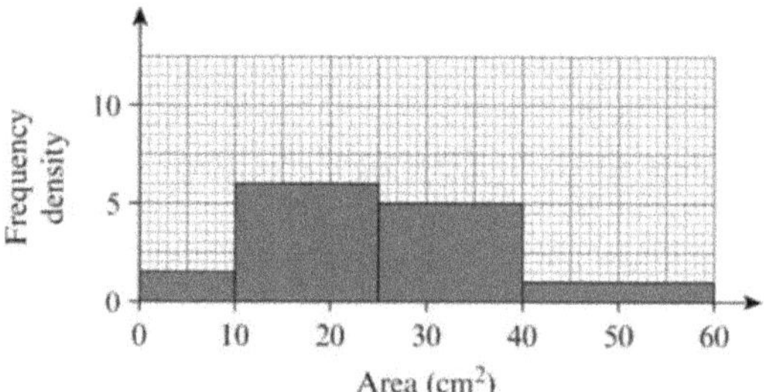

Question 3

The table shows the time taken by students to solve a puzzle.

Time (t seconds)	Frequency
$0 < t \leq 10$	8
$10 < t \leq 20$	16
$20 < t \leq 25$	15
$25 < t \leq 30$	12
$30 < t \leq 50$	6

Use this information to complete the histogram.

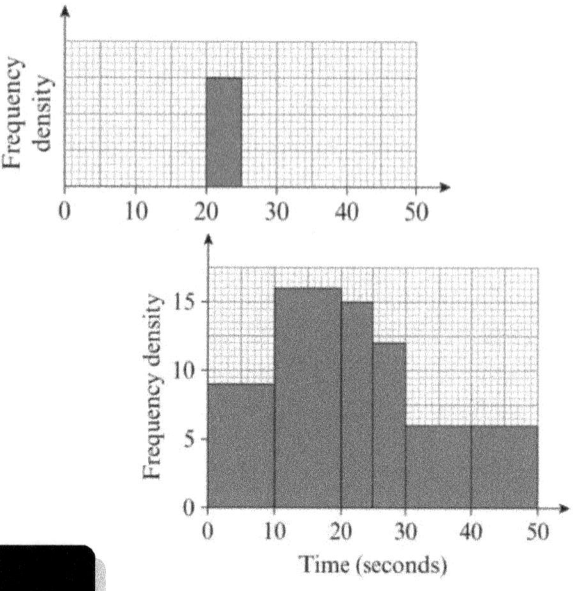

Question 4

The histogram shows information about the lifetime of batteries.

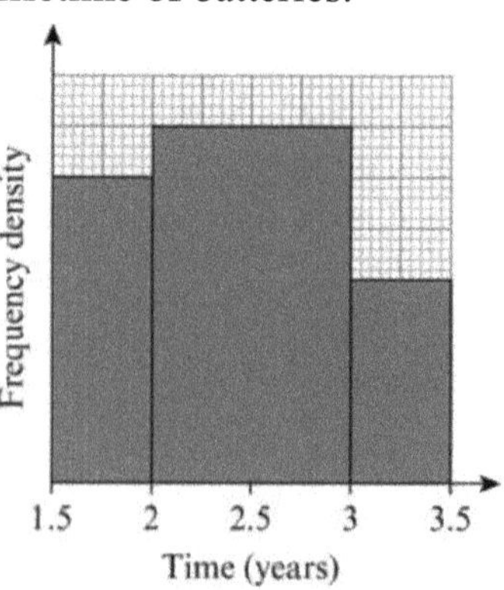

15 of the batteries had a lifetime between 1.5 and 2 years. Work out the total number of batteries shown on the histogram.

$15 + 20 + 10 = 45$

Answers

Question 1

They have just added up the heights (frequency densities) of the bars and not the frequencies.

The correct answer is

Total number of cases = 10 + 15 + 18 + 20 + 18 = 81

$\frac{38}{81} \times 100 = 46.9\,\%$ to 3 significant figures.

Question 2

They have made a mistake in calculating the last frequency density and they have misread the scale.

The correct working is:

Area (A cm^2)	Frequency	CW	FD
$0 < A \le 10$	30	10	3
$10 < A \le 25$	35	5	7
$25 < A \le 40$	75	15	5
$40 < A \le 60$	20	20	1

The correct histogram is:

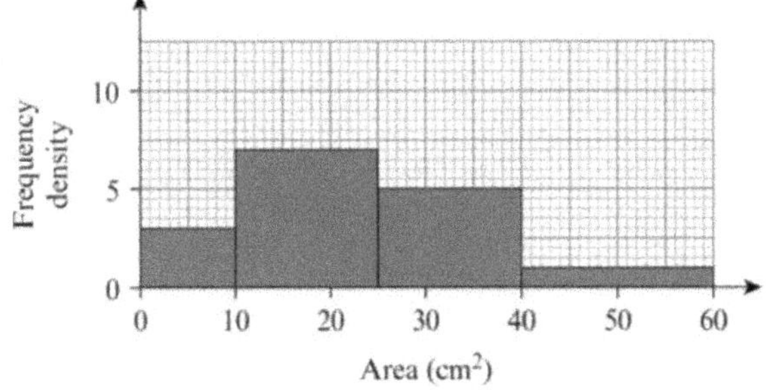

Question 3

They have just plotted frequencies on the vertical axis instead of frequency densities.

The correct working is

Time (t seconds)	Frequency	CW	FD
$0 < t \le 10$	8	10	0.8
$10 < t \le 20$	16	10	1.6
$20 < t \le 25$	15	5	3
$25 < t \le 30$	12	5	2.4
$30 < t \le 50$	6	20	0.3

The correct histogram is

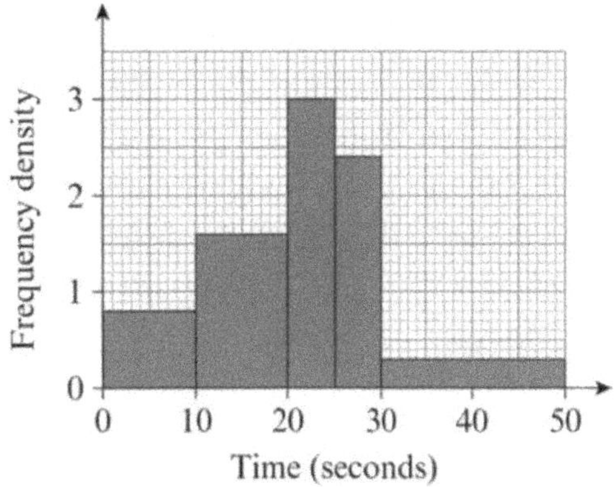

Question 4

They have just used frequencies instead of frequency density on the vertical axis.

The correct answer is:

Number of years	1.5 to 2	2 to 3	3 to 3.5
Frequency	15	35	10
Frequency density	30	35	20

Total = 15 + 35 + 10 = 60

Solving simultaneous equations card sort

Teacher notes

Many students struggle to solve simultaneous equations. This can be because they can only solve simultaneous equations by elimination. This is a good method, but there are alternatives which may require less work and have less chance of error. Some students forget to rearrange simultaneous equations before attempting to solve them, or are unable to rearrange them accurately.

This activity is suitable as a plenary or revision activity after the work on solving simultaneous equations algebraically. It can also be used as an alternative to a student book exercise. Students need to understand the difference between solving simultaneous equations by elimination, substitution and equating expressions (a type of substitution), and to be able to rearrange equations accurately.

Introductory activity

You may want to revise the three methods for solving simultaneous equations. The purpose of this card sort activity is for students to decide which method is best in each case: encourage them to comment on the relative merits of each method.

Method 1: solving by elimination

$4x - 5y = 4$	(1)	multiply (1) by 2 to give (3)
$6x + 2y = 25$	(2)	multiply (2) by 5 to give (4)
$8x - 10y = 8$	(3)	
$30x + 10y = 125$	(4)	add (3) to (4) to **eliminate** y
$38x = 133$		solve to find x
$x = 3.5$		substitute $x = 3.5$ into (2)
$6 \times 3.5 + 2y = 25$		
$21 + 2y = 25$		solve to find y
$2y = 4$		
$y = 2$		

Method 2: solving by substitution

$2x + y = 1$	(1)	rearrange (1) to make y the subject to give (3)
$3x + 4y = 6$	(2)	
$y = 1 - 2x$	(3)	substitute (3) into (2)
$3x + 4(1 - 2x) = 6$		solve to find x
$3x + 4 - 8x = 6$		
$-5x = 2$		
$x = -0.4$		substitute $x = -0.4$ into (3)
$y = 1 - 2 \times -0.4$		
$y = 1.8$		

Method 3: solving by equating the expressions

(This is a version of the substitution method.)

$y = 2 - x$	(1)	
$y = 2x - 4$	(2)	equate the expressions
$2 - x = 2x - 4$		solve to find x
$6 = 3x$		
$x = 2$		substitute $x = 2$ into (1)
$y = 2 - 2$		
$y = 0$		

Card sorting activity

Ask students to work in small groups. Print out the sort cards and cut them up. Ask students to sort the cards into three piles according to what they think is the best method of solving them. The three categories are: **elimination**, **substitution** and **equating the expressions**. Once students have sorted the cards into the three piles you could ask them to explain when each particular method is best. For example, they may say:

Equating the expressions method is best when the coefficient of either x or y is 1 or -1 for **both** equations.

Substitution method is best when the coefficient of either x or y is 1 or -1 for **one** equation.

Elimination method is best when the coefficient of either x or y is not 1 or -1 for **either** equation.

Differentiation

In IGCSE maths students are not expected to solve simultaneous equations where one is linear and one is quadratic but this is in the 'A' level maths syllabus. To stretch your most able students you can include the final two cards to provide additional discussion.

Alternative approaches

You could ask students to decide which equations are easier to solve and which are harder and why. For example, they may say:

In the equating the expressions method it is easier when x (or y) is the subject of both equations and you don't have to rearrange the expressions first.

In the substitution method it is easier when x (or y) is the subject of one equation and so you don't have to rearrange the expression.

In the elimination method it is easier when you do not have to multiply any expression by anything, and you can just add or subtract the equations as they are, or you only have to multiply one of the equations by something.

You could extend the activity by asking students to pick what they think are the hardest pair to solve from each category and ask them to solve them.

Answers

There is a case to be argued for equations to go in different columns. Students should be able to justify their decisions, for example $x + 7y = 54$ and $x + 2y = 19$ could go in any of the three columns!

Equating the expressions		**Substitution**		**Elimination**	
Equations	**Solution**	**Equations**	**Solution**	**Equations**	**Solution**
$y = 10 + x$ $y = 7x - 2$	$x = 2$ $y = 12$	$x = y - 2$ $2x + 3y = 21$	$x = 3$ $y = 5$	$3x - 4y = -17$ $2x - 2y = -8$	$x = 1$ $y = 5$
$y = x + 8$ $3x - 6 = y$	$x = 7$ $y = 15$	$y = 2x + 1$ $3y - 4x = 7$	$x = 2$ $y = 5$	$4x + 3y = 21$ $3 = 3x - 2y$	$x = 3$ $y = 3$
$x = 20 - y$ $x = 2y + 8$	$x = 16$ $y = 4$	$r = 3v - 1$ $v + 2r = 12$	$r = 5$ $v = 2$	$3x = 30 - 2y$ $6x + 3y = 57$	$x = 8$ $y = 3$
$x = y + 8$ $4 - 3y = x$	$x = 7$ $y = -1$	$2b - 7a = 11$ $6b + a = -11$	$a = -2$ $b = -1.5$	$3p - 4m = 21$ $5m = 2p - 17.5$	$p = 5$ $m = -1.5$
$p = -2m - 1$ $p = 3 - 10m$	$p = -2$ $m = 0.5$	$x - y - 7 = 0$ $3x - 2y = 18$	$x = 4$ $y = -3$	$2p + 3g + 3 = 5$ $3p + 2g - 2 = -9$	$p = -5$ $g = 4$
$t = -3 - 16r$ $t = 4r - 8$	$t = -7$ $r = 0.25$	$5 - x = 2y$ $3x + 5y - 11 = 0$	$x = -3$ $y = 4$	$2x + 3y - 7 = 0$ $3x - 4y - 2 = 0$	$x = 2$ $y = 1$
				$0.5 = 5h + 7f$ $-29 + 6f = 10h$	$h = -2$ $f = 1.5$
				$5x - 8 = 4y$ $6x - 5y - 10 = 0$	$x = 0$ $y = -2$
				$2p = 23 + 5w$ $3w - 4p - 1 = -19$	$p = 1.5$ $w = -4$
Could go in any of the three columns:		$x + 7y = 54$ $x + 2y = 19$	$x = 5$ $y = 7$		
		$x + 2y = 3$ $x^2 + 3xy = 10$	$x = 4$ 5 $y = -0.5$ -1		
		$2x + y = 1$ $x^2 + y^2 = 1$	$x = 0$ 0.8 $y = 1$ -0.6		

Note: the extra examples, which are **not** on the IGCSE® syllabus, are solved by substitution and require students to be able to solve a quadratic equation; each pair of equations has two solutions.

Solving simultaneous equations sort cards

$3x - 4y = -17$ $2x - 2y = -8$	$y = 2x + 1$ $3y - 4x = 7$
$4x + 3y = 21$ $3 = 3x - 2y$	$x + 7y = 54$ $x + 2y = 19$
$x = 20 - y$ $x = 2y + 8$	$y = x + 8$ $3x - 6 = y$
$5 - x = 2y$ $3x + 5y - 11 = 0$	$x - y - 7 = 0$ $3x - 2y = 18$
$x = y - 2$ $2x + 3y = 21$	$2p + 3g + 3 = 5$ $3p + 2g - 2 = -9$
$t = -3 - 16r$ $t = 4r - 8$	$p = -2m - 1$ $p = 3 - 10m$
$2x + 3y - 7 = 0$ $3x - 4y - 2 = 0$	$0.5 = 5h + 7f$ $-29 + 6f = 10h$
$r = 3v - 1$ $v + 2r = 12$	$x = y + 8$ $4 - 3y = x$
$3x = 30 - 2y$ $6x + 3y = 57$	$y = 10 + x$ $y = 7x - 2$
$5x - 8 = 4y$ $6x - 5y - 10 = 0$	$3p - 4m = 21$ $5m = 2p - 17.5$
$2b - 7a = 11$ $6b + a = -11$	$2p = 23 + 5w$ $3w - 4p - 1 = -19$
$x + 2y = 3$ $x^2 + 3xy = 10$	$2x + y = 1$ $x^2 + y^2 = 1$

Graphs of inequalities and linear programming collective memory

Teacher notes

This activity is a revision activity on graphing inequalities and linear programming. For a collective memory activity students are shown a poster for a short period of time, and they must try to reproduce as much of the poster as they can. There is an opportunity to highlight common misconceptions, encourage teamwork and make maths lessons, and in particular note taking or revising, more interesting and fun.

Introductory activity

There should be little need for any introduction, as this is intended as a revision exercise.

Collective memory activity

You will need to have on your desk a copy of the collective memory poster (preferably double the size, A3, using the colour version from the online resources). You will need a stopwatch, (you can find one online at http://www.online-stopwatch.com/). Divide the class into groups of four students. Give each group a blank sheet of A3 paper and coloured pens/pencils to match those used in the memory poster. Tell students that during this activity you expect them to communicate with their teammates and devise strategies for how to approach this task.

Give each student in the group a number from 1 to 4. To start all students numbered 1 come to the front of the class and have 30 seconds to view the poster. Then they return to their group and have one minute to produce as much of the poster as they can remember, as well as communicate with their teammates what areas the next person should concentrate on. Then all students numbered 2 view the poster for 30 seconds and have one minute back with their team. This continues until each team member has seen the poster twice. Allow two minutes at the end for the group to finish off their poster.

At the end of the activity reveal the poster to the class and discuss the activity considering some of the following points.

- Which parts of the poster were easier to recreate, which were more difficult and why?
- What are the most important features of the poster?
- How well did they work as a team?

Differentiation

To make this task harder do not allow the student who viewed the poster to draw anything, instead the next numbered person in the team has to do the drawing. This means even more cooperation and communication is required.

You could make it easier or harder by:

- adjusting the time for which students are allowed to view or draw
- allowing more or fewer than two views per student
- putting more or less information in the poster.

Alternative approaches

You could display an electronic copy of the memory poster on the board to avoid the need for students to move around the classroom and the entire group can view the poster at the same time, which can speed up the activity. You can still keep the rule that only one student can write at a time and that they must take it in turns to do so.

You could ask students to work individually instead of in groups.

Answers

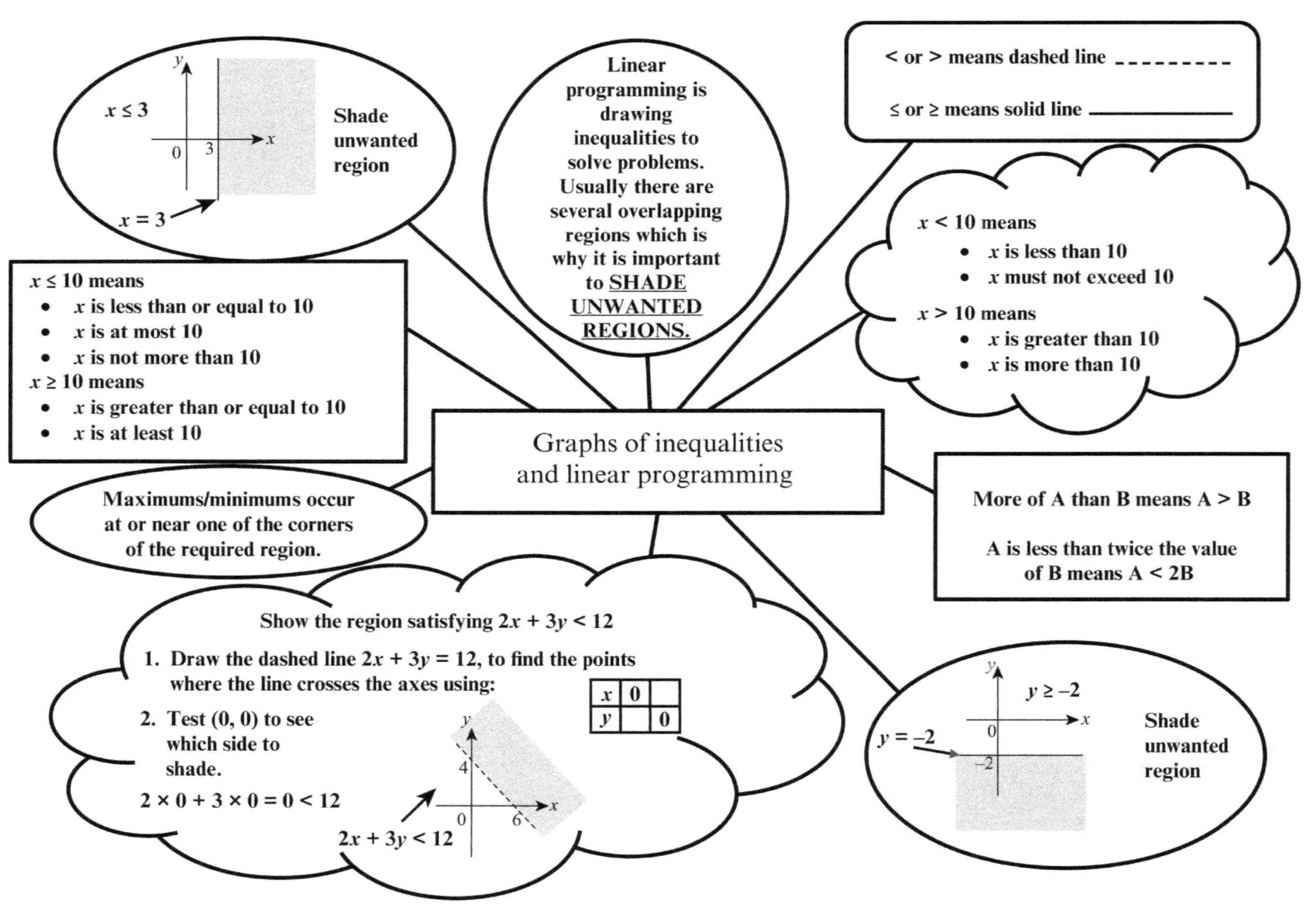

Graphs of inequalities and linear programming
< or > means dashed line
≤ or ≥ means solid line
Linear programming is drawing inequalities to solve problems. Usually there are several overlapping regions which is why it is important to SHADE UNWANTED REGIONS.
x ≤ 3
0
3
x
y
x = 3
Shade unwanted region
x < 10 means
• x is less than 10
• x must not exceed 10
x > 10 means
• x is greater than 10
• x is more than 10
x ≤ 10 means
• x is less than or equal to 10
• x is at most 10
• x is not more than 10
x ≥ 10 means
• x is greater than or equal to 10
• x is at least 10
Maximums/minimums occur at or near one of the corners of the required region.
More of A than B means A > B
A is less than twice the value of B means A < 2B
Show the region satisfying 2x + 3y < 12
1. Draw the dashed line 2x + 3y = 12, to find the points where the line crosses the axes using:
x | 0 |
y | | 0
2. Test (0, 0) to see which side to shade.
2 × 0 + 3 × 0 = 0 < 12
2x + 3y < 12
4
6
0
y ≥ −2
y = −2
−2
Shade unwanted region

Quadratic graphs (factorising and completing the square) card game

Teacher notes

For quadratic functions, being able to complete the square, factorise, understand the purpose of these processes and the connection with the graph are very useful. To complete this activity, students should have already learned how to factorise and that factorising helps find the solutions to the equations. These solutions are where the curve crosses the x-axis. Students should also be able to complete the square, as this is an alternative way to solve equations.

Introductory activity

Explain to students that there are four cards in a set; a graph (not drawn to scale), a factorised expression, an equation and the completed square version.

For example, a set might contain:

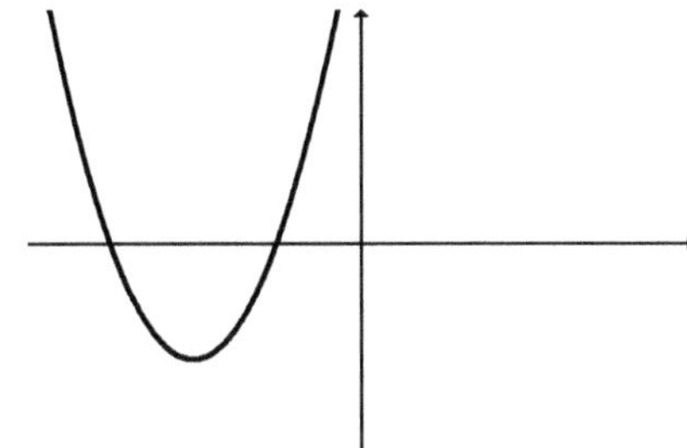

$y = x^2 + 6x + 5$

$y = (x + 1)(x + 5)$

Completing the square:

$y = (x + 3)^2 - 4$

Students need to look at the diagrams very carefully; if both solutions are negative (as in the above example $x = -1$ or $x = -5$) the curve will cross the x-axis twice on the left hand side of the y-axis.

Card game

To play the game students should work in small groups of two to four players. Each group will need one complete set of cards (eleven pages photocopied and cut up to make 80 cards).

The aim is for players to make two sets of three matching cards (they do not need all four cards to make a matching set).

One player should shuffle the cards and then deal six cards to each player; the remaining cards are placed on the table face down in a pile.

Turn the top card face up and place it on the discard pile next to the face-down cards.

Players take it in turns to draw a card from the face-down pile before they discard a card they don't want.

Play continues to the next player in a clockwise direction each taking it in turns to draw then discard a card. Play finishes when a player has two complete sets of three cards. The other players must check they are matching sets.

Note: diagrams can go with more than one set of equations. Students will hopefully decide (or can be told) that these are the worst cards to discard for that reason.

Differentiation

To make it a little easier you could include the "wild cards", which can replace any ONE card in a set of three.

To make it harder one player should deal eight cards to each player and then the remaining cards are placed on the table face down in a pile. This time the aim is for players to make two sets of four matching cards.

You could discuss how completing the square helps us find the coordinates of the maximum/ minimum point, for example, $y = (x - p)^2 + q$ has a minimum point at (p, q). This is not required on the IGCSE syllabus but is on the 'A' level syllabus so this could be useful to extend the more able.

Alternative approaches

You could use the cards to play a game of snap in pairs. Each player is dealt half the cards then they both turn one card over face up simultaneously. If they are in the same set they shout snap. If they shout snap first they win those cards. The winner is the player with all the cards at the end or more cards after a set amount of time.

You could use several copies of the blank cards from the online resources to ask students to make their own set. In all the printed sets the coefficient of x^2 is 1, students could make a set where the coefficient of x^2 is not 1.

Answers

The correct sets are grouped in rows below – keep one set as an answer sheet.

Quadratic graphs game cards

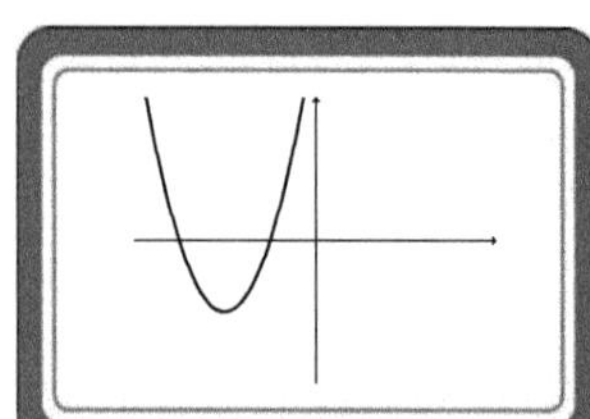	$y = x^2 + 8x + 12$	$y = (x + 2)(x + 6)$	Completing the square: $y = (x + 4)^2 - 4$
	$y = x^2 + 10x + 21$	$y = (x + 3)(x + 7)$	Completing the square: $y = (x + 5)^2 - 4$
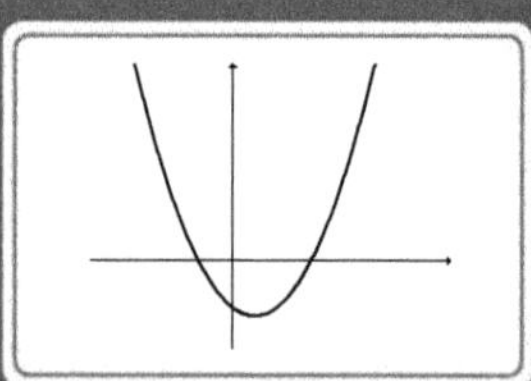	$y = x^2 - 2x - 15$	$y = (x + 3)(x - 5)$	Completing the square: $y = (x - 1)^2 - 16$
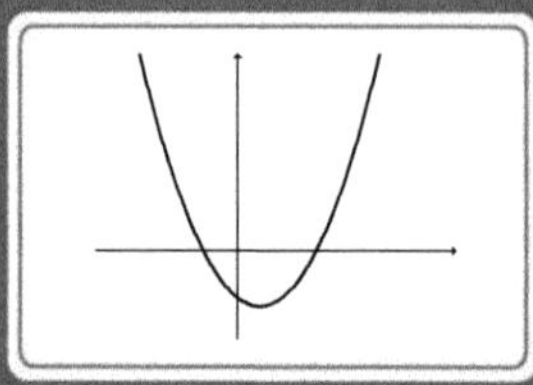	$y = x^2 - 4x - 12$	$y = (x + 2)(x - 6)$	Completing the square: $y = (x - 2)^2 - 16$
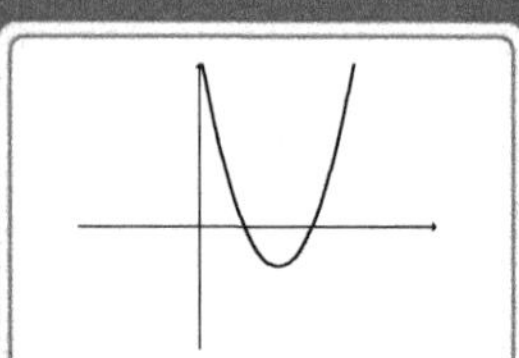	$y = x^2 - 6x + 5$	$y = (x - 1)(x - 5)$	Completing the square: $y = (x - 3)^2 - 4$

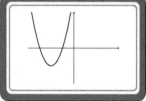

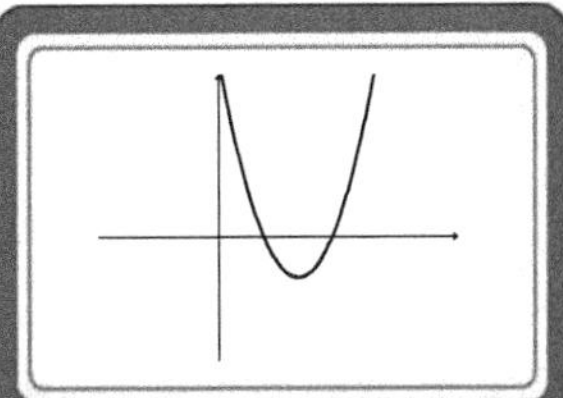	$y = x^2 - 12x + 27$	$y = (x - 3)(x - 9)$	Completing the square: $y = (x - 6)^2 - 9$
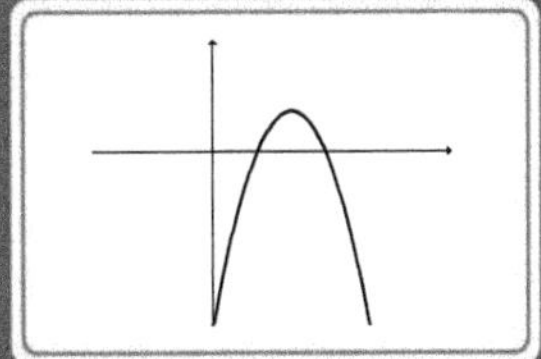	$y = -x^2 + 8x - 15$	$y = (3 - x)(x - 5)$	Completing the square: $y = -(x - 4)^2 + 1$
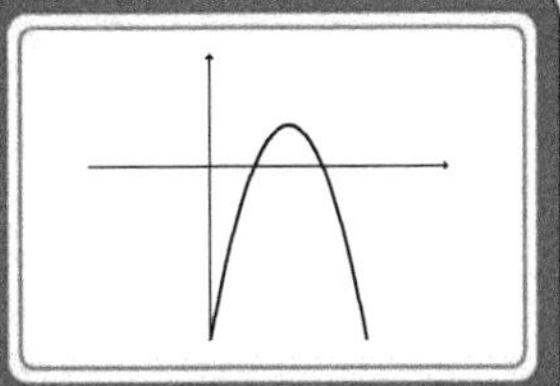	$y = -x^2 + 10x - 21$	$y = (3 - x)(x - 7)$	Completing the square: $y = -(x - 5)^2 + 4$
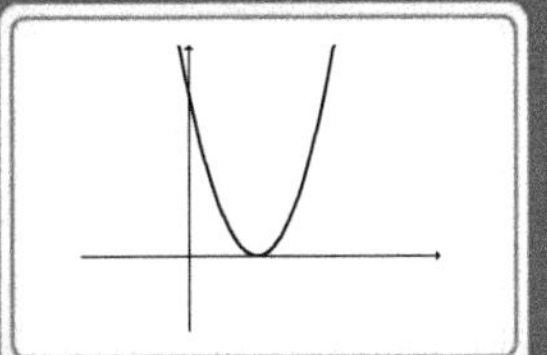	$y = x^2 - 10x + 25$	$y = (x - 5)^2$	Completing the square: $y = (x - 5)^2$
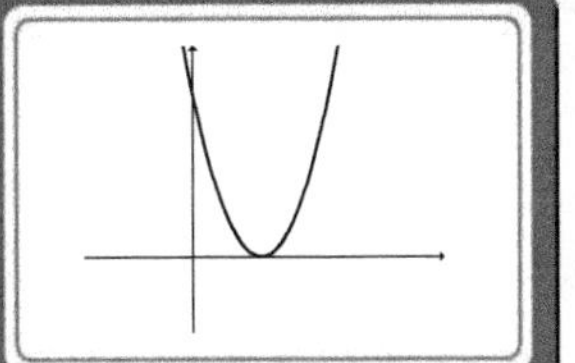	$y = x^2 - 4x + 4$	$y = (x - 2)^2$	Completing the square: $y = (x - 2)^2$
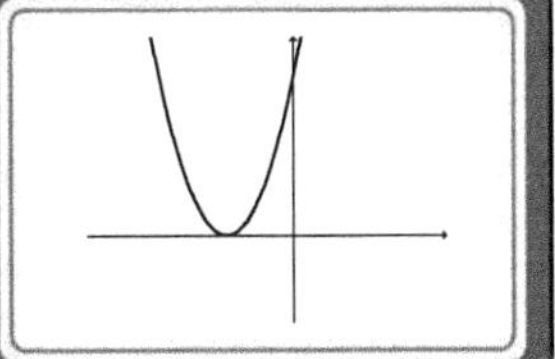	$y = x^2 + 6x + 9$	$y = (x + 3)^2$	Completing the square: $y = (x + 3)^2$
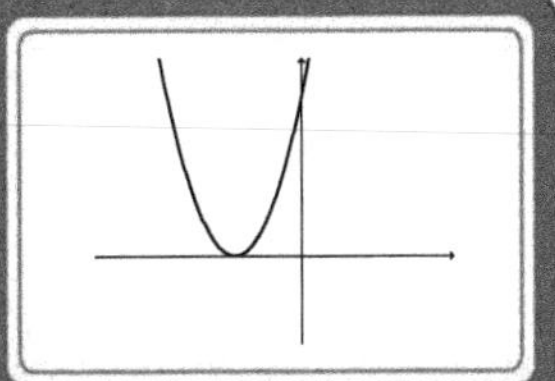	$y = x^2 + 14x + 49$	$y = (x + 7)^2$	Completing the square: $y = (x + 7)^2$
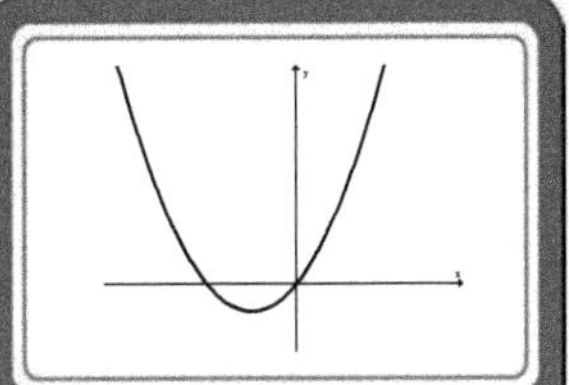	$y = x^2 + 6x$	$y = x(x + 6)$	Completing the square: $y = (x + 3)^2 - 9$

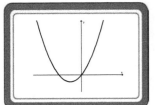

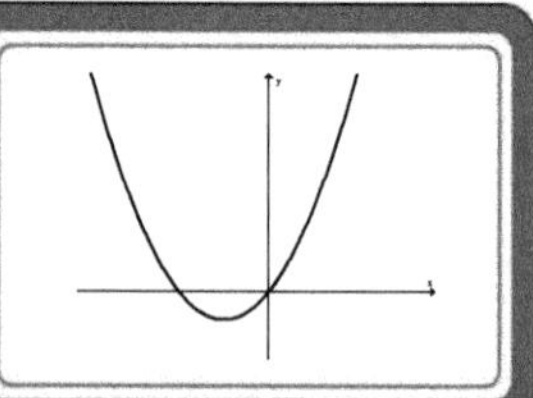	$y = x^2 + 12x$	$y = x(x + 12)$	Completing the square: $y = (x + 6)^2 - 36$
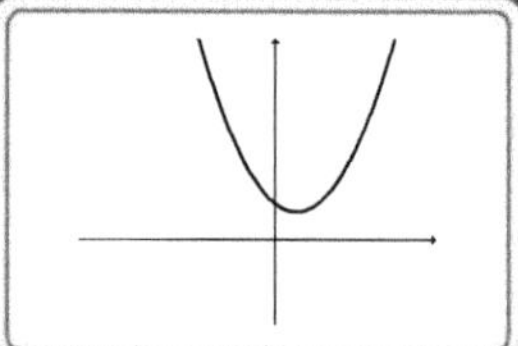	$y = x^2 - 2x + 17$	Will not factorise	Completing the square: $y = (x - 1)^2 + 16$
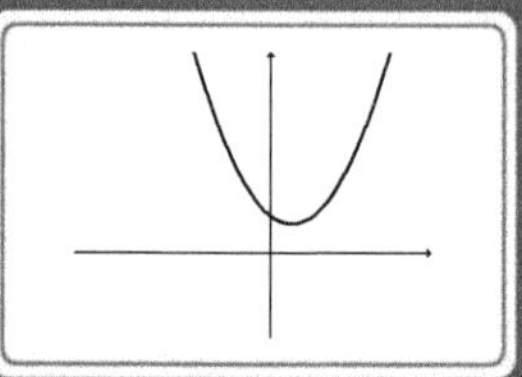	$y = x^2 - 6x + 12$	Will not factorise	Completing the square: $y = (x - 3)^2 + 3$
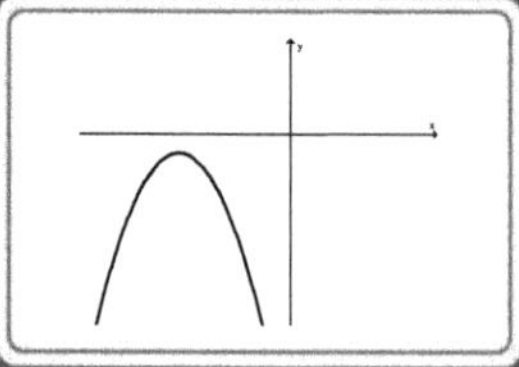	$y = -x^2 - 8x - 17$	Will not factorise	Completing the square: $y = -(x + 4)^2 - 1$
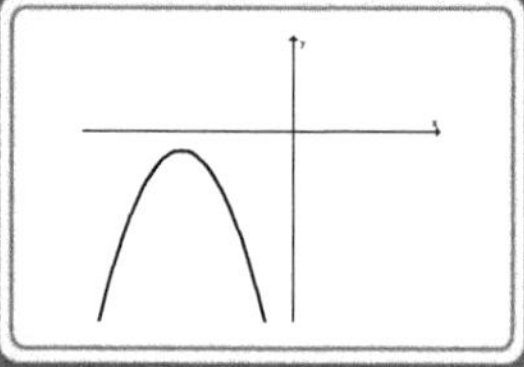	$y = -x^2 - 4x - 6$	Will not factorise	Completing the square: $y = -(x + 2)^2 - 2$
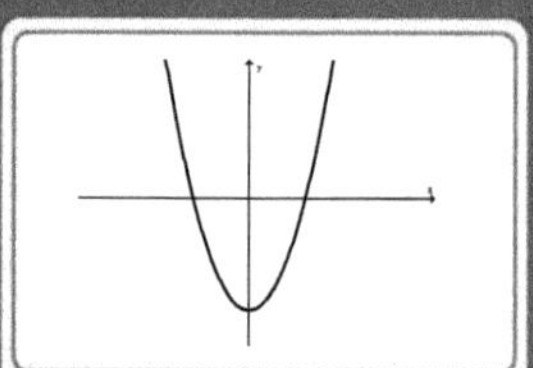	$y = x^2 - 25$	$y = (x + 5)(x - 5)$	Completing the square: $y = x^2 - 25$
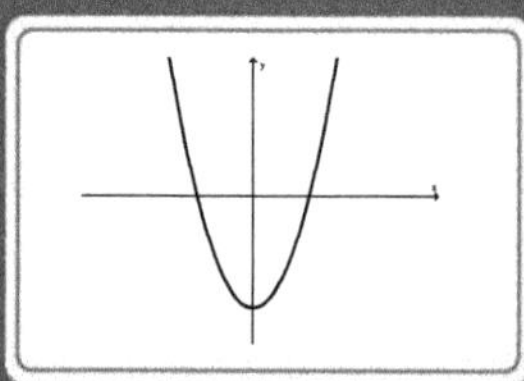	$y = x^2 - 81$	$y = (x + 9)(x - 9)$	Completing the square: $y = x^2 - 81$
		WILD CARD	WILD CARD

Larger size jigsaw pieces are available in the online resources.

Teacher notes

This activity can be used as a starter, a plenary, a homework task or a revision exercise. This fun task revises the basics of vectors work, in particular drawing vectors, adding and subtracting them and multiplying them by a scalar.

Introductory activity

This is quite a basic exercise however you may wish to revise the following.

Example 1

Subtracting vectors:

$$\begin{pmatrix} 5 \\ -2 \end{pmatrix} - \begin{pmatrix} 2 \\ 1 \end{pmatrix} = \begin{pmatrix} 3 \\ -3 \end{pmatrix}$$

Example 2

Multiplying by a scalar:

$$3\begin{pmatrix} 5 \\ -2 \end{pmatrix} = \begin{pmatrix} 15 \\ -6 \end{pmatrix}$$

The direction of the vector remains the same but it is three times as long.

Example 3

Combining adding and multiplication by a scalar:

If $\mathbf{a} = \begin{pmatrix} 2 \\ 1 \end{pmatrix}$ and $\mathbf{b} = \begin{pmatrix} 4 \\ -3 \end{pmatrix}$ find $\frac{1}{2}(\mathbf{a}+\mathbf{b})$

$$\frac{1}{2}\left[\begin{pmatrix} 2 \\ 1 \end{pmatrix} + \begin{pmatrix} 4 \\ -3 \end{pmatrix}\right] = \frac{1}{2}\begin{pmatrix} 6 \\ -2 \end{pmatrix} = \begin{pmatrix} 3 \\ -1 \end{pmatrix}$$

Example 4

Negative vectors:

If $\mathbf{c} = \begin{pmatrix} 2 \\ -5 \end{pmatrix}$ then $-\mathbf{c} = \begin{pmatrix} -2 \\ 5 \end{pmatrix}$

The vector remains the same length but the direction is reversed.

Example 5

Drawing vectors:

Starting at the X, draw the vector $\begin{pmatrix} 2 \\ -3 \end{pmatrix}$ followed by $\begin{pmatrix} 6 \\ 0 \end{pmatrix}$ then $\begin{pmatrix} -5 \\ 2 \end{pmatrix}$ and finally $\begin{pmatrix} 0 \\ 1 \end{pmatrix}$.

Start each new vector where the previous one finished.

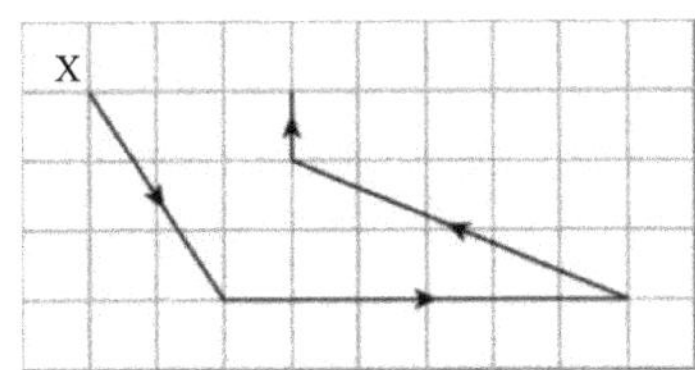

Vector picture

Give each student a copy of the sheet. Ask them to draw the vectors from the bottom of the sheet in order starting on the grid at the letter X. There is no need for them to draw the direction arrows on the vectors for this activity since these are not necessary for their picture.

Differentiation

You could make the task harder by asking students to devise alternatives for some of the vectors. For example,

vector $\mathbf{h} = \begin{pmatrix} -4 \\ 1 \end{pmatrix}$ could be replaced with $-4\mathbf{i} + \mathbf{j}$ and

vector $\mathbf{f} = \begin{pmatrix} -1 \\ -2 \end{pmatrix}$ could be replaced with $-\frac{1}{2}\mathbf{d}$

Alternative approaches

Ask the students to make their own vector picture and set of instructions for drawing it. A blank sheet for this is provided in the online resources. Ask students to consider why they think vectors **i** and **j** have been left in on the blank sheet.

Answers

The picture is a shark

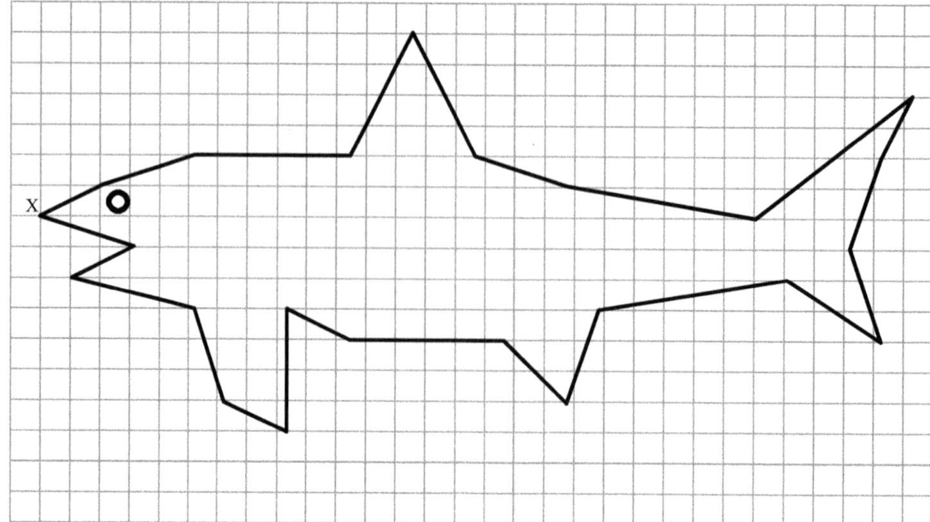

1 $\mathbf{a} = \begin{pmatrix} 2 \\ 1 \end{pmatrix}$

2 $\mathbf{b} = \begin{pmatrix} 3 \\ 1 \end{pmatrix}$

3 $\mathbf{c} = \begin{pmatrix} 5 \\ 0 \end{pmatrix}$

4 $\mathbf{d} = \begin{pmatrix} 2 \\ 4 \end{pmatrix}$

5 $2\mathbf{i} - 4\mathbf{j} = 2\begin{pmatrix} 1 \\ 0 \end{pmatrix} - 4\begin{pmatrix} 0 \\ 1 \end{pmatrix} = \begin{pmatrix} 2 \\ 0 \end{pmatrix} - \begin{pmatrix} 0 \\ 4 \end{pmatrix} = \begin{pmatrix} 2 \\ -4 \end{pmatrix}$

6 $\mathbf{e} = \begin{pmatrix} 3 \\ -1 \end{pmatrix}$

7 $2\mathbf{b} - 3\mathbf{j} = 2\begin{pmatrix} 3 \\ 1 \end{pmatrix} - 3\begin{pmatrix} 0 \\ 1 \end{pmatrix} = \begin{pmatrix} 6 \\ 2 \end{pmatrix} - \begin{pmatrix} 0 \\ 3 \end{pmatrix} = \begin{pmatrix} 6 \\ -1 \end{pmatrix}$

8 $\mathbf{a} + \mathbf{b} + 2\mathbf{j} = \begin{pmatrix} 2 \\ 1 \end{pmatrix} + \begin{pmatrix} 3 \\ 1 \end{pmatrix} + 2\begin{pmatrix} 0 \\ 1 \end{pmatrix}$

$= \begin{pmatrix} 5 \\ 2 \end{pmatrix} + \begin{pmatrix} 0 \\ 2 \end{pmatrix} = \begin{pmatrix} 5 \\ 4 \end{pmatrix}$

9 $\mathbf{f} = \begin{pmatrix} -1 \\ -2 \end{pmatrix}$

10 $\mathbf{g} - 2\mathbf{i} = \begin{pmatrix} 1 \\ -3 \end{pmatrix} - 2\begin{pmatrix} 1 \\ 0 \end{pmatrix} = \begin{pmatrix} 1 \\ -3 \end{pmatrix} - \begin{pmatrix} 2 \\ 0 \end{pmatrix} = \begin{pmatrix} -1 \\ -3 \end{pmatrix}$

11 $\mathbf{g} = \begin{pmatrix} 1 \\ -3 \end{pmatrix}$

12 $\mathbf{j} - \mathbf{e} = \begin{pmatrix} 0 \\ 1 \end{pmatrix} - \begin{pmatrix} 3 \\ -1 \end{pmatrix} = \begin{pmatrix} -3 \\ 2 \end{pmatrix}$

13 $-3\mathbf{j} - 2\mathbf{e} = -3\begin{pmatrix} 0 \\ 1 \end{pmatrix} - 2\begin{pmatrix} 3 \\ -1 \end{pmatrix}$

$= \begin{pmatrix} 0 \\ -3 \end{pmatrix} - \begin{pmatrix} 6 \\ -2 \end{pmatrix} = \begin{pmatrix} -6 \\ -1 \end{pmatrix}$

14 $\mathbf{g} - 2\mathbf{i} = \begin{pmatrix} 1 \\ -3 \end{pmatrix} - 2\begin{pmatrix} 1 \\ 0 \end{pmatrix} = \begin{pmatrix} 1 \\ -3 \end{pmatrix} - \begin{pmatrix} 2 \\ 0 \end{pmatrix} = \begin{pmatrix} -1 \\ -3 \end{pmatrix}$

15 $\mathbf{b} - \mathbf{c} + \mathbf{j} = \begin{pmatrix} 3 \\ 1 \end{pmatrix} - \begin{pmatrix} 5 \\ 0 \end{pmatrix} + \begin{pmatrix} 0 \\ 1 \end{pmatrix}$

$= \begin{pmatrix} -2 \\ 1 \end{pmatrix} + \begin{pmatrix} 0 \\ 1 \end{pmatrix} = \begin{pmatrix} -2 \\ 2 \end{pmatrix}$

16 $-\mathbf{c} = -\begin{pmatrix} 5 \\ 0 \end{pmatrix} = \begin{pmatrix} -5 \\ 0 \end{pmatrix}$

17 $\frac{1}{2}(\mathbf{h} + \mathbf{j}) = \frac{1}{2}\left[\begin{pmatrix} -4 \\ 1 \end{pmatrix} + \begin{pmatrix} 0 \\ 1 \end{pmatrix}\right] = \frac{1}{2}\begin{pmatrix} -4 \\ 2 \end{pmatrix} = \begin{pmatrix} -2 \\ 1 \end{pmatrix}$

18 $-4\mathbf{j} = -4\begin{pmatrix} 0 \\ 1 \end{pmatrix} = \begin{pmatrix} 0 \\ -4 \end{pmatrix}$

19 $\mathbf{f} - \mathbf{g} = \begin{pmatrix} -1 \\ -2 \end{pmatrix} - \begin{pmatrix} 1 \\ -3 \end{pmatrix} = \begin{pmatrix} -2 \\ 1 \end{pmatrix}$

20 $5\mathbf{j} - \frac{1}{2}\mathbf{d} = 5\begin{pmatrix} 0 \\ 1 \end{pmatrix} - \frac{1}{2}\begin{pmatrix} 2 \\ 4 \end{pmatrix} = \begin{pmatrix} 0 \\ 5 \end{pmatrix} - \begin{pmatrix} 1 \\ 2 \end{pmatrix} = \begin{pmatrix} -1 \\ 3 \end{pmatrix}$

21 $\mathbf{h} = \begin{pmatrix} -4 \\ 1 \end{pmatrix}$

22 $\mathbf{a} = \begin{pmatrix} 2 \\ 1 \end{pmatrix}$

23 $-\mathbf{e} = -\begin{pmatrix} 3 \\ -1 \end{pmatrix} = \begin{pmatrix} -3 \\ 1 \end{pmatrix}$

Vectors

Using the following vectors and grid

$\mathbf{a} = \begin{pmatrix} 2 \\ 1 \end{pmatrix}$ $\mathbf{b} = \begin{pmatrix} 3 \\ 1 \end{pmatrix}$ $\mathbf{c} = \begin{pmatrix} 5 \\ 0 \end{pmatrix}$ $\mathbf{d} = \begin{pmatrix} 2 \\ 4 \end{pmatrix}$ $\mathbf{e} = \begin{pmatrix} 3 \\ -1 \end{pmatrix}$

$\mathbf{f} = \begin{pmatrix} -1 \\ -2 \end{pmatrix}$ $\mathbf{g} = \begin{pmatrix} 1 \\ -3 \end{pmatrix}$ $\mathbf{h} = \begin{pmatrix} -4 \\ 1 \end{pmatrix}$ $\mathbf{i} = \begin{pmatrix} 1 \\ 0 \end{pmatrix}$ $\mathbf{j} = \begin{pmatrix} 0 \\ 1 \end{pmatrix}$

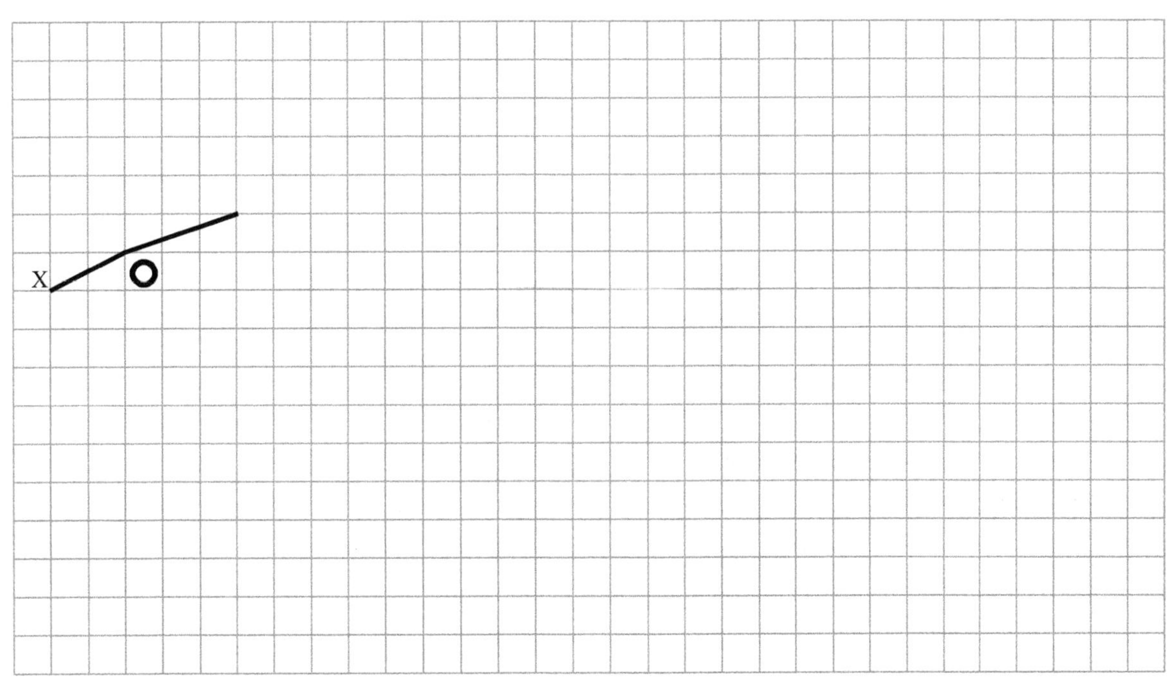

start at the X and draw the **following** vectors in order.
The first two are drawn for you. Start each new vector
where the previous one finishes. What picture do you get?

USE A PENCIL NOT A PEN.

1	$\mathbf{a}$	**2**	$\mathbf{b}$	**3**	$\mathbf{c}$	**4**	$\mathbf{d}$
5	$2\mathbf{i} - 4\mathbf{j}$	**6**	$\mathbf{e}$	**7**	$2\mathbf{b} - 3\mathbf{j}$	**8**	$\mathbf{a} + \mathbf{b} + 2\mathbf{j}$
9	$\mathbf{f}$	**10**	$\mathbf{g} - 2\mathbf{i}$	**11**	$\mathbf{g}$	**12**	$\mathbf{j} - \mathbf{e}$
13	$-3\mathbf{j} - 2\mathbf{e}$	**14**	$\mathbf{g} - 2\mathbf{i}$	**15**	$\mathbf{b} - \mathbf{c} + \mathbf{j}$	**16**	$-\mathbf{c}$
17	$\frac{1}{2}(\mathbf{h} + \mathbf{j})$	**18**	$-4\mathbf{j}$	**19**	$\mathbf{f} - \mathbf{g}$	**20**	$5\mathbf{j} - \frac{1}{2}\mathbf{d}$
21	$\mathbf{h}$	**22**	$\mathbf{a}$	**23**	$-\mathbf{e}$		

Upper and lower bounds revision wheel

Teacher notes

It is important that students can confidently calculate the upper and lower bounds of given numbers, before beginning work on finding appropriate upper and lower bounds to solutions of simple problems. This activity is a jigsaw-type activity with only one solution. Each question involves asking for either the upper or lower bound to a given rounded number or the range of possible values for that rounded number in the form of an inequality.

This is an interesting alternative approach to using a written exercise as it promotes discussion, allows for group work, and the students often find it enjoyable. Students can do the questions in any order and can leave questions that they find difficult until the end. You and the students will know when you have finished as the puzzle will make a complete circle.

Introductory activity

If necessary you could give students the following example as an introduction.

6.84 rounded to 2 d.p. has a lower bound of 6.835 and an upper bound of 6.845

Its true value lies between 6.835 and 6.845 which we can write as the following inequality
$6.835 \leq n < 6.845$

Jigsaw activity

Print out the jigsaw pieces and cut them up. There are 24 pieces in a complete set. Ask the students to work in small groups to complete the jigsaw. There is only one correct solution.

Differentiation

You could give students a time limit for this activity, or challenge the students to see which group can complete the activity first. You could leave out two or three cards in a section and ask students to make their own cards in order to complete the wheel. Blank cards are available in the online resources.

Alternative approaches

You could enlarge the jigsaw pieces to do one large jigsaw as a class or ask students to be creative to complete their own version (blank jigsaw pieces are available in the online resources).

Answers

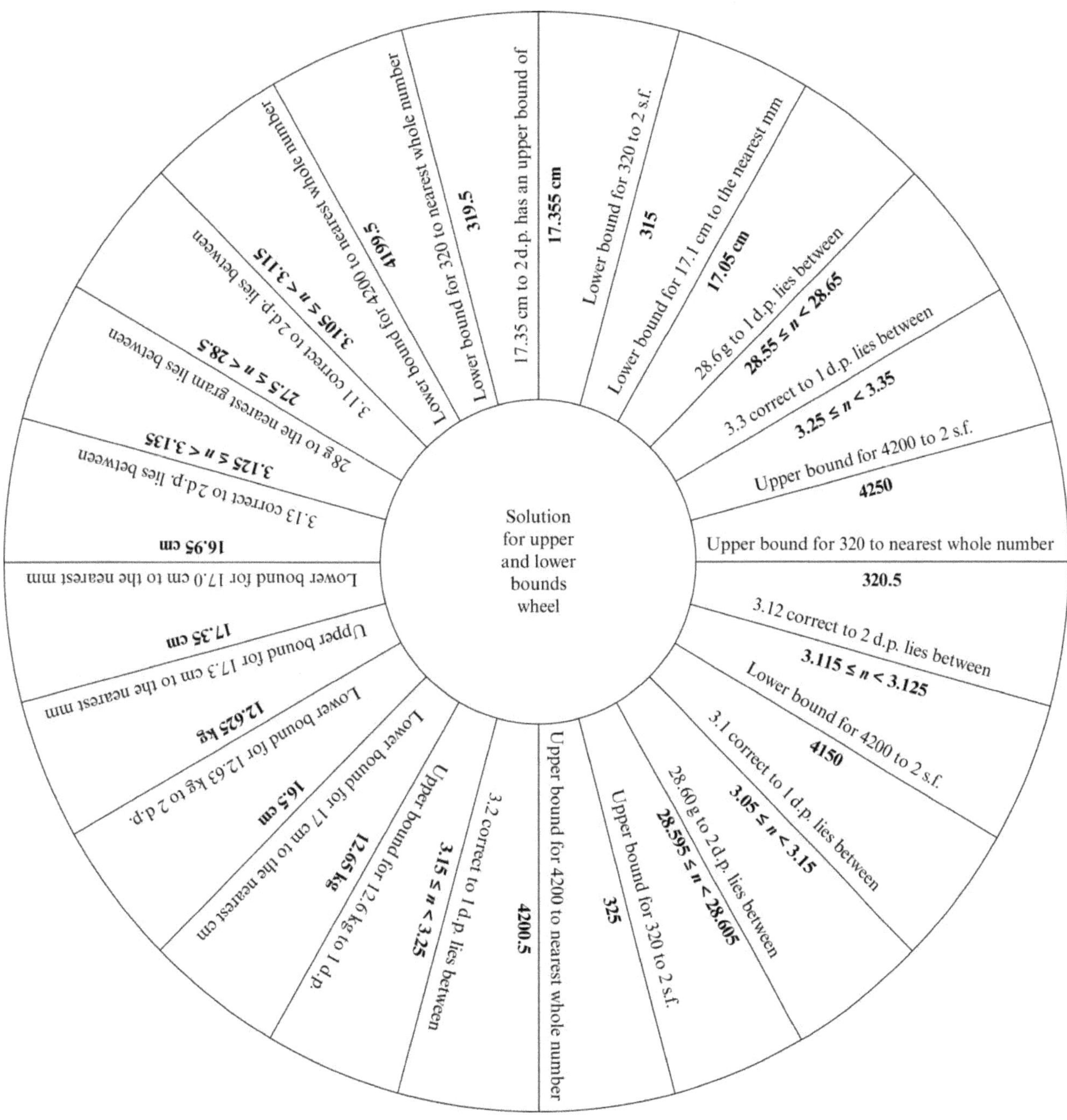
Solution for upper and lower bounds wheel
17.35 cm to 2 d.p. has an upper bound of
17.355 cm
Lower bound for 320 to 2 s.f.
315
Lower bound for 17.1 cm to the nearest mm
17.05 cm
28.6 g to 1 d.p. lies between
28.55 ≤ n < 28.65
3.3 correct to 1 d.p. lies between
3.25 ≤ n < 3.35
Upper bound for 4200 to 2 s.f.
4250
Upper bound for 320 to nearest whole number
320.5
3.12 correct to 2 d.p. lies between
3.115 ≤ n < 3.125
Lower bound for 4200 to 2 s.f.
4150
3.1 correct to 1 d.p. lies between
3.05 ≤ n < 3.15
28.60 g to 2 d.p. lies between
28.595 ≤ n < 28.605
Upper bound for 320 to 2 s.f.
325
Upper bound for 4200 to nearest whole number
4200.5
3.2 correct to 1 d.p. lies between
3.15 ≤ n < 3.25
Upper bound for 12.6 kg to 1 d.p.
12.65 kg
Lower bound for 17 cm to the nearest cm
16.5 cm
Lower bound for 12.63 kg to 2 d.p.
12.625 kg
Upper bound for 17.3 cm to the nearest mm
17.35 cm
Lower bound for 17.0 cm to the nearest mm
16.95 cm
3.13 correct to 2 d.p. lies between
3.125 ≤ n < 3.135
28 g to the nearest gram lies between
27.5 ≤ n < 28.5
3.11 correct to 2 d.p. lies between
3.105 ≤ n < 3.115
Lower bound for 4200 to nearest whole number
4199.5
Lower bound for 320 to nearest whole number
319.5

Upper and lower bound revision wheel jigsaw pieces

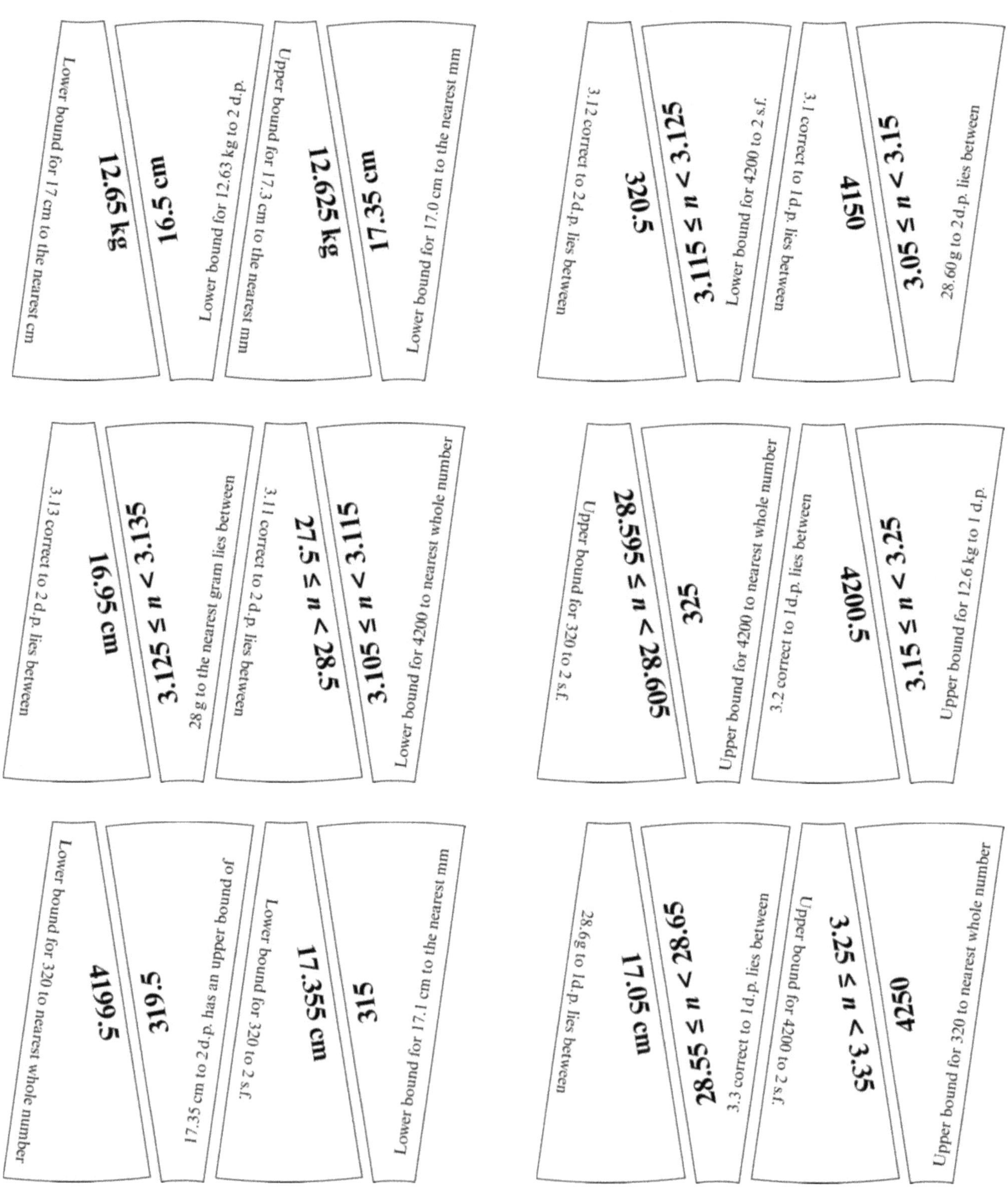

Larger size jigsaw pieces are available in the online resources.

Differentiation dominoes

Teacher notes

This dominoes activity is an interesting and enjoyable alternative to a written exercise, and allows students to discuss their ideas. If they are not sure how to do a particular question, students can choose an alternative starting point. This activity requires students to be able to understand the idea of a derived function, to then be able to find the derivatives of functions of the form $f(x) = ax^n$ and simple sums of these.

They need to apply differentiation to gradients and turning points (stationary points) and discriminate between maxima and minima by any method.

Introductory activity

If this activity is used as a revision activity or a plenary you may find that only a little introduction is necessary; otherwise you may want to go through the following:

Example 1

Understand the different notations that may be used:

a) $y = 3x^2$ then $\frac{dy}{dx} = 6x$

b) $\frac{d}{dx}\left(x^4 - x\right) = 4x^3 - 1$

c) $f(x) = 5x\left(2 - 3x^2\right)$; expand the brackets first to get $f(x) = 10x - 15x^3$ then $f'(x) = 10 - 45x^2$

Example 2

Applying differentiation:

a) Find the gradient of the curve $y = \left(3x - 2\right)^2$ at the point (3, 49).

Expand the brackets to get $y = 9x^2 - 12x + 4$

Differentiate to get $\frac{dy}{dx} = 18x - 12$

Substitute $x = 3$ into $\frac{dy}{dx}$ to get $18 \times 3 - 12 = 42$

b) The curve $y = x^3 - 9x^2 + 15x - 8$ has two turning points. Find the minimum turning point.

First differentiate $\frac{dy}{dx} = 3x^2 - 18x + 15$

Turning points occur when $\frac{dy}{dx} = 0$ so

$3x^2 - 18x + 15 = 0$

Factorise to solve $(3x - 15)(x - 1) = 0$ gives turning points at $x = 1$ and $x = 5$.

Find the second derivative $\frac{d^2y}{dx^2} = 6x - 18$

At $x = 1$ then $\frac{d^2y}{dx^2} = 6 \times 1 - 18 = -12$. As this is negative, there is a maximum at $x = 1$.

At $x = 5$ then $\frac{d^2y}{dx^2} = 6 \times 5 - 18 = 12$. As this is positive, there is a minimum at $x = 5$.

Find the y-coordinate to go with $x = 5$ by substituting into $y = x^3 - 9x^2 + 15x - 8$

$y = 5^3 - 9 \times 5^2 + 15 \times 5 - 8 = -33$ so the minimum point is at (5, –33)

Domino activity

Ask your students to work in small groups. Copy or print out the dominoes onto card and cut them up into individual rectangles, each of which consists of two squares. Note they will *not* be pairs where the two expressions on the same domino are equal. For example, the first domino is:

$y = x^2 + 5x$	$\frac{dy}{dx} = 10x$ Find $\frac{d^2y}{dx^2}$

Give each group a complete set of domino pieces. (It might be useful to do each set on a different coloured piece of card. If any piece is dropped or lost, you will know what set it is from). Ask students to join the dominoes together to form a chain so that adjacent expressions match.

Differentiation

To make it harder you could remove one or two of the pieces from somewhere in the chain; and replace them with blank dominoes. Students should then write the correct expressions on these blank pieces in order to fill in the gaps.

Alternative approaches

Using the blank domino pieces from the online resources, you could ask students to use these pieces to make an entirely new domino activity of their own. Alternatively, ask them to consider extending this chain by replacing the piece that says 'Finish' on it.

Answers

Start
$\frac{dy}{dx} = 2x - 10$

$f(x) = 4x^3 - 8x$ Find $f'(3)$	10	$\frac{dy}{dx} = 10x$ Find $\frac{d^2y}{dx^2}$	$y = x^2 + 5x$	$\frac{dy}{dx} = 2x + 5$	$y = x - 5^2$

100
Find the coordinates of the maximum point on $y = -2(x^3 - 9x^2 + 24x - 16)$

(4, 0)	64	$f(x) = x^5 - 2x^4$ Find $f''(2)$	Find the coordinates of the minimum point on $y = 2x^3 - 6x$	(1, −4)	(2, −7)

Answers (continued)

Find the coordinates of the minimum point on $y=2x^2-8x+1$
102

$y=10x-3$	$\frac{d}{dx}(x^4+4x^2-5)$	$4x^3+8x$	$\frac{dy}{dx}10x^2-3$	$y=5x^2-3x$	Find the value of $\frac{dy}{dx}$ for $y=\frac{(9x-2)^2}{6}$ at $x=4$

$\frac{dy}{dx}=10$
$\frac{d}{dx}((x^2-2)^2)$

$4x^3-8x$	$\frac{d}{dx}(4x(1-x))$	$4-8x$	$f(x)=$ $(x-3)(x+4)$ Find $f'(50)$	101	Finish

Differentiation dominoes

$y=x^2+5x$	$\frac{dy}{dx}=10x$ Find $\frac{d^2y}{dx^2}$	Find the coordinates of the minimum point on $y=2x^2-8x+1$	102
$f(x)=x^5-2x^4$ Find $f''(2)$	Find the coordinates of the minimum point on $y=2x^3-6x$	10	$f(x)=4x^3-8x$ Find $f'(3)$
$y=x-5^2$	$\frac{dy}{dx}=2x+5$	$\frac{d}{dx}(x^4+4x^2-5)$	$y=10x-3$
Find the value of $\frac{dy}{dx}$ for $y=\frac{(9x-2)^2}{6}$ at $x=4$	$y=5x^2-3x$	(4, 0)	64
100	Find the coordinates of the maximum point on $y=-2(x^3-9x^2+$ $24x-16)$	$\frac{dy}{dx}=10$	$\frac{d}{dx}((x^2-2)^2)$
$4-8x$	$f(x)=(x-3)(x+4)$ Find $f'(50)$	$\frac{dy}{dx}10x^2-3$	$4x^3+8x$
(1, –4)	(2, –7)	101	Finish
$4x^3-8x$	$\frac{d}{dx}(4x(1-x))$	Start	$\frac{dy}{dx}=2x-10$

Expanding more than two brackets jigsaw

Teacher notes

This jigsaw activity is an interesting group work alternative to a written exercise. The advantages of a jigsaw activity over an exercise are that it creates opportunities for students to discuss their ideas, and they can find an alternative starting point if they are not sure how to do a particular question. This activity requires students to be able to expand more than two brackets.

Introductory activity

If you want to use this task as a revision activity or plenary you may find that only a little introduction is necessary. Otherwise you may want to go through the following example.

Example

Expand $(3x-2)^2(x+4)$

First expand and simplify two of the brackets
$(3x-2)(3x-2)(x+4) = (9x^2-6x-6x+4)(x+4)$

Then expand the result $(9x^2-12x+4)(x+4)$
$= 9x^3+36x^2-12x^2-48x+4x+16$

Finally simplify $9x^3+36x^2-12x^2-48x+4x+16$
$= 9x^3+24x^2-44x+16$

Jigsaw activity

Ask your students to work in small groups. Copy all the jigsaw pieces onto card to make a complete set of jigsaw pieces for each group. There are several different-shaped jigsaws in the Teacher Resource Pack so you should not tell students what shape to expect. This jigsaw does not have any blank edges and this means that it is harder for your students to work out where the edges are. It is useful to make each set a different colour. If any piece is dropped or lost, you will know what set it is from.

Differentiation

To make it more difficult, there are questions or 'solutions' that should end up around the edge of the jigsaw, which are associated with common errors made in expanding brackets. If students make such an error in expanding another bracket in the jigsaw, they may be tempted to think one of these 'edge pieces' is the answer. If you want to make it easier, you could indicate the outside edges: perhaps draw a bold line along the outside edges.

Alternative approaches

You could just ask students to complete the jigsaw, or you could make use of the fact that the outside edges are not blank. You could print blank jigsaw pieces from the online resources and ask students to use these pieces to extend the jigsaw, either finding the solutions to the harder questions or, if possible, writing a question that has the given answer. Alternatively, students could make their own, entirely new jigsaw.

Answers

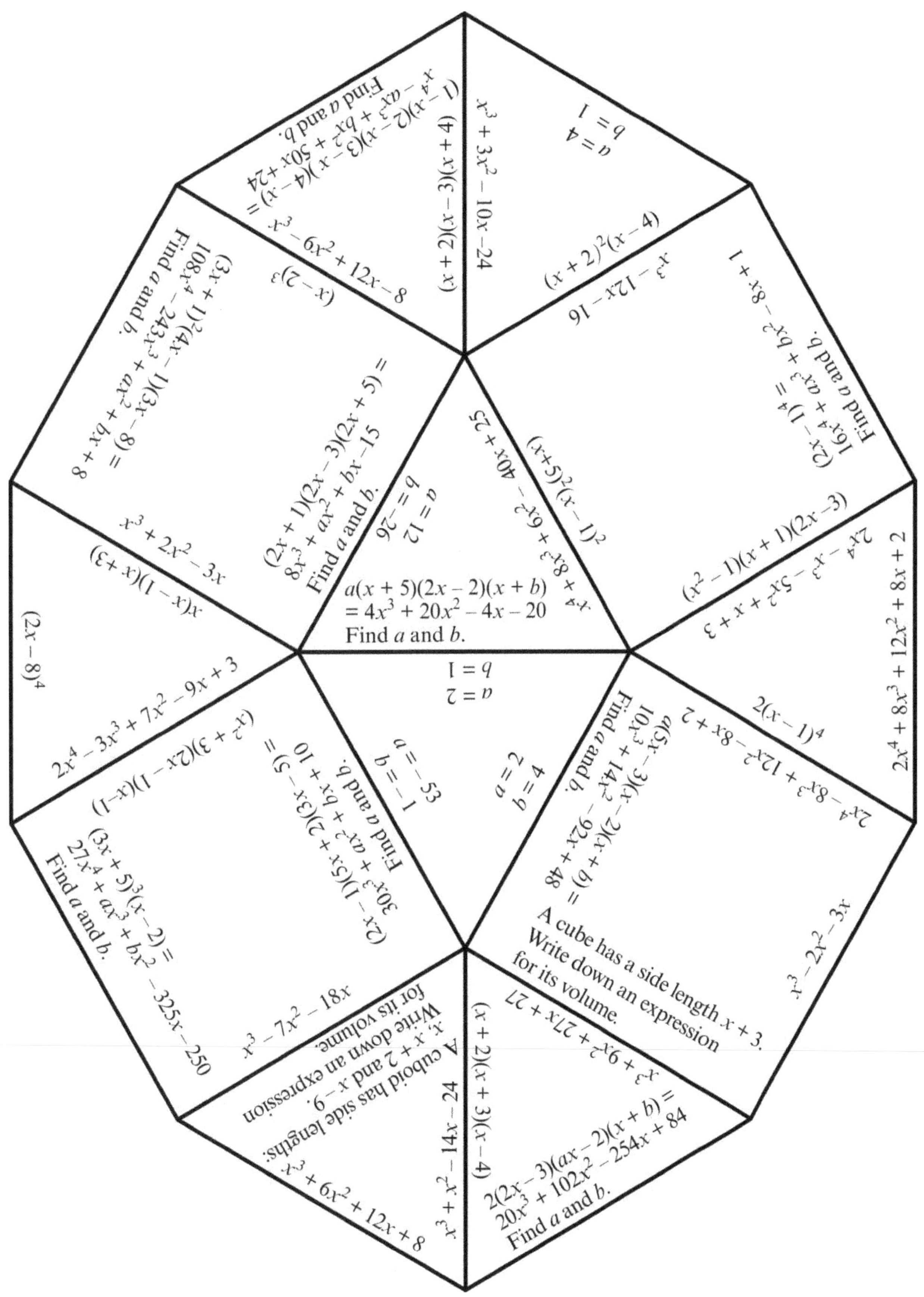

Expanding more than two brackets jigsaw pieces

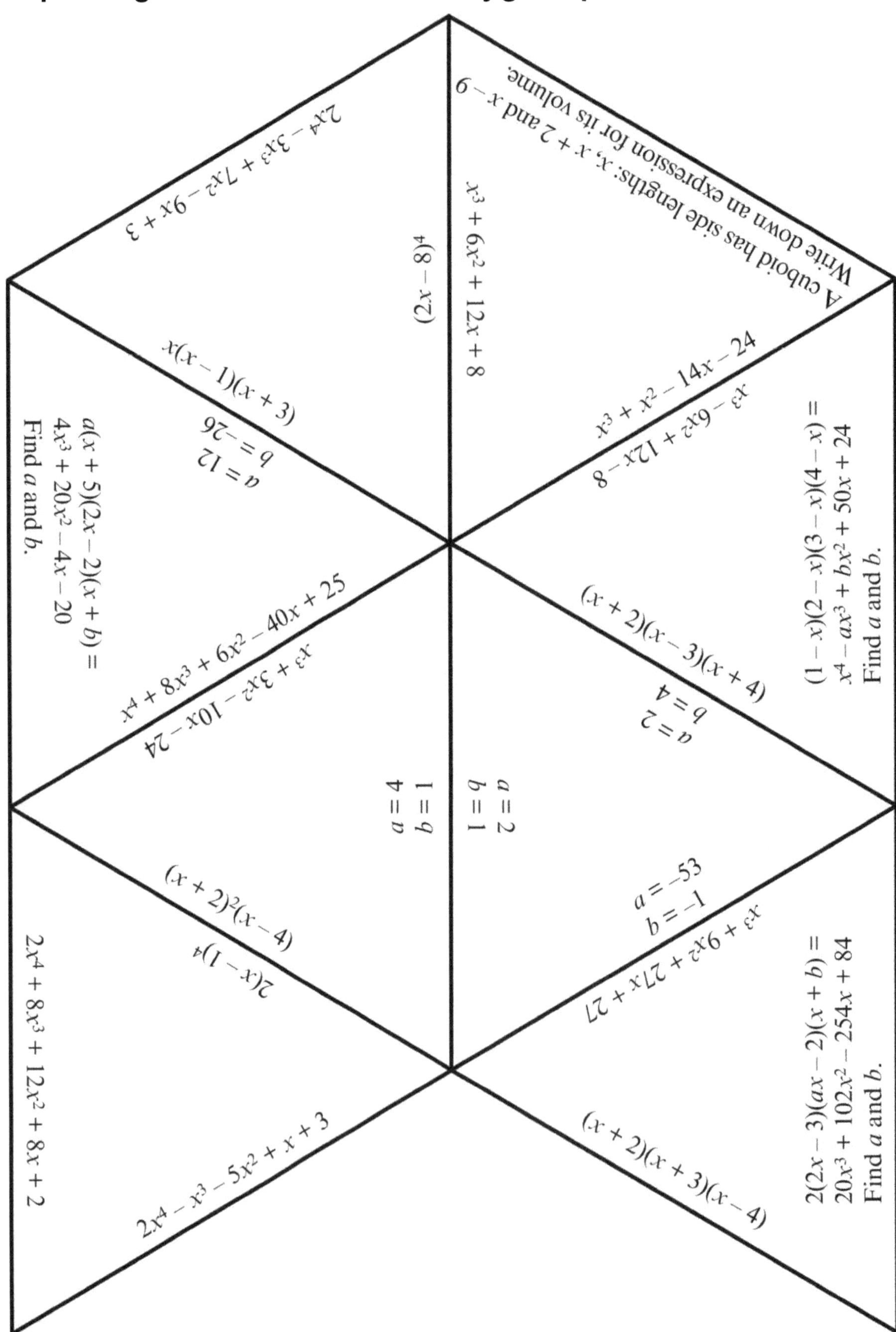

A cuboid has side lengths: x, $x + 2$ and $x - 9$
Write down an expression for its volume.
$2x^4 - 3x^3 + 7x^2 - 9x + 3$
$(2x - 8)^4$
$x^3 + 6x^2 + 12x + 8$
$x(x - 1)(x + 3)$
$a = 12$
$b = -26$
$x^3 + x^2 - 14x - 24$
$x^3 - 6x^2 + 12x - 8$
$a(x + 5)(2x - 2)(x + b) =$
$4x^3 + 20x^2 - 4x - 20$
Find a and b.
$(1 - x)(2 - x)(3 - x)(4 - x) =$
$x^4 - ax^3 + bx^2 + 50x + 24$
Find a and b.
$x^4 + 8x^3 + 6x^2 - 40x + 25$
$x^3 + 3x^2 - 10x - 24$
$(x + 2)(x - 3)(x + 4)$
$a = 2$
$b = 4$
$a = 4$
$b = 1$
$a = 2$
$b = 1$
$(x + 2)^2(x - 4)$
$2(x - 1)^4$
$a = -53$
$b = -1$
$x^3 + 9x^2 + 27x + 27$
$2x^4 + 8x^3 + 12x^2 + 8x + 2$
$2x^4 - x^3 - 5x^2 + x + 3$
$(x + 2)(x + 3)(x - 4)$
$2(2x - 3)(ax - 2)(x + b) =$
$20x^3 + 102x^2 - 254x + 84$
Find a and b.

$(2x + 1)(2x - 3)(2x + 5) =$
$8x^3 + ax^2 + bx - 15$
Find a and b.

$x^3 + 2x^2 - 3x$

$(x - 2)^3$

$(3x + 1)^2(4x - 1)(3x - 8) =$
$108x^4 - 243x^3 + ax^2 + bx + 8$
Find a and b.

A cube has a side length $x + 3$
Write down an expression for its volume.

$a(5x - 3)(x - 2)(x + b) =$
$10x^3 + 14x^2 - 92x + 48$
Find a and b.

$x^3 - 2x^2 - 3x$

$2x^4 - 8x^3 + 12x^2 - 8x + 2$

$x^3 - 12x - 16$

$(x + 5)^2(x - 1)^2$

$(2x - 1)^4 =$
$16x^4 + ax^3 + bx^2 - 8x + 1$
Find a and b.

$(x^2 - 1)(x + 1)(2x - 3)$

$(3x + 5)^3(x - 2) =$
$27x^4 + ax^3 + bx^2 - 325x - 250$
Find a and b.

$x^3 - 7x^2 - 18x$

$(x^2 + 3)(2x - 1)(x - 1)$

$(2x - 1)(5x + 2)(3x - 5) =$
$30x^3 + ax^2 + bx + 10$
Find a and b.

Stem-and-leaf Venn diagram sorting

Teacher notes

This task can be used as a plenary or a revision activity. It is an interesting task as it revises a lot of topics at once. The task covers interpretation of stem-and-leaf diagrams, as well as revision of averages, range and Venn diagrams. Students will be able to revise the work using group discussion.

Introductory activity

If the activity is used as a revision tool at the end of Unit 10, then very little introduction on interpreting stem-and-leaf diagrams should be necessary. It is probably worth a quick recap of averages, range and Venn diagrams.

5	3 4 8
6	1 2 4 7
7	0 2 2 4
8	3 6
9	1
10	1

Key

5|3 means 5.3

To find the median from this stem-and-leaf diagram with 15 ordered leaves, find the 8th value (use $\frac{15+1}{2}$) which is 7.0.

To find the mode, find the most frequently recurring value: 7.2.

To find the range, find the highest value and subtract the lowest value: $10.1 - 5.3 = 4.8$.

To find the mean, find the sum of all values and then divide by 15. In this case, the mean is 7.12.

An example of a Venn diagram to sort the integers between 1 and 17 is:

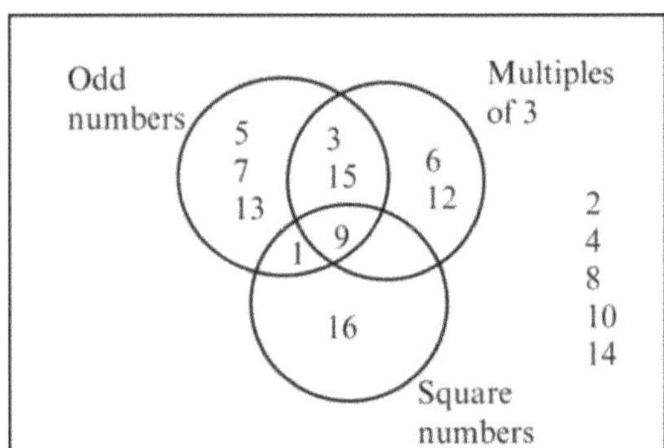

Talk about the **eight** different regions in the Venn diagram particularly highlighting the overlapping regions and the region outside the three circles.

Ask students for the smallest number that could go in the blank region if you extended the numbers used. (Answer 36).

Venn diagram sorting activity

On card, print out the stem-and-leaf diagrams labelled A to H. Ask your students to work in small groups. Give each group a set of the eight stem-and-leaf diagrams and a copy of the Venn diagram sheet. Ask students to sort each of the stem-and-leaf diagrams into the correct region on the Venn diagram.

Differentiation

To make this activity easier you could omit one of the circles and just use the top two circles of the Venn diagram.

To make this activity harder you could miss out one or two of the stem-and-leaf diagrams and ask the students to fill in the gaps with their own diagrams. You could also give extra stem-and-leaf diagrams to sort. You could also give them a time limit for the activity.

Alternative approaches

Using blank stem-and-leaf cards (templates below), ask students to create their own stem-and-leaf diagrams for each of the regions. This is a useful extra activity for early finishers.

Alternatively, you could ask students to think of different criteria for the regions on the Venn diagram, and then sort the stem-and-leaf diagrams A to H accordingly. Is there a region in their Venn diagram that is not possible?

Blank cards

1 |
2 |
3 |
4 |
5 |

Key

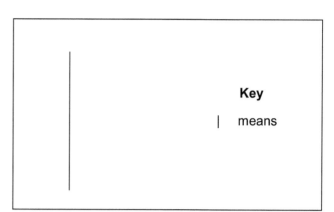

1| means

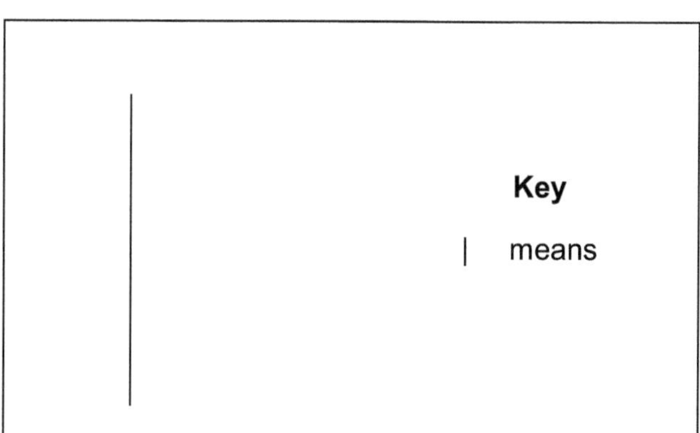

Answers

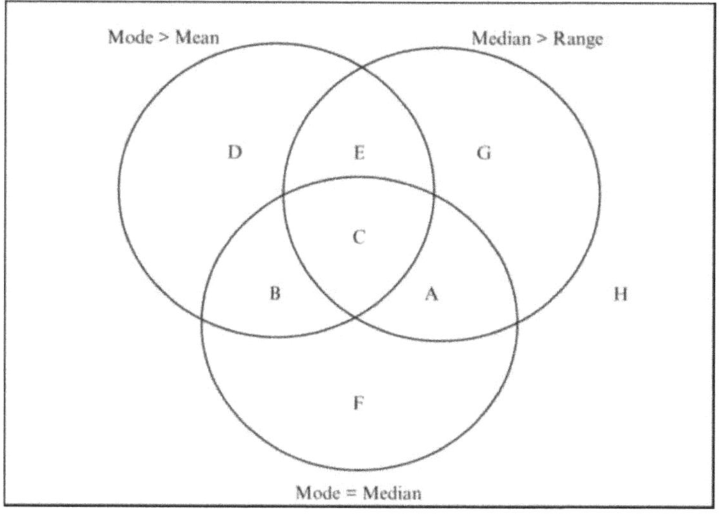

Statistics

	Set A	Set B	Set C	Set D	Set E	Set F	Set G	Set H
Mean	35.6	35.2	36.6	28.6	27.2	31	39.6	39.6
Mode	35	36	39	31	42	31	23	21
Median	35	36	39	28	26	31	42	42
Range	34	40	33	44	25	44	39	48

Stem and leaf sorting cards

A

Stem	Leaf
1	8
2	0 1 7
3	1 2 5 5
4	0 1 2 3 7
5	0 2

Key
1|8 means 18

B

Stem	Leaf
1	2
2	0 1 7
3	1 2 6 6
4	0 1 2 3 5
5	0 2

Key
1|2 means 12

C

Stem	Leaf
1	8 9
2	0 6
3	1 2 9 9
4	0 4 5 7 8
5	0 1

Key
1|8 means 18

D

Stem	Leaf
1	0 2 3 9
2	0 3 6 8
3	0 1 1 9
4	5 8
5	4

Key
1|0 means 10

E

Stem	Leaf
1	7 8 9
2	0 1 2 3 6 8 9
3	1 2 8
4	2 2

Key
1|7 means 17

F

Stem	Leaf
1	0 7 9
2	3 6 7 9
3	1 1 1 1 9
4	8 9
5	4

Key
1|0 means 10

G

Stem	Leaf
1	9
2	3 3 6
3	0 1 9
4	2 4 8 9
5	3 4 5 8

Key
1|9 means 19

H

Stem	Leaf
1	1
2	1 1 3
3	0 1 9
4	2 8 9
5	3 4 5 8 9

Key
1|1 means 11

Venn diagram

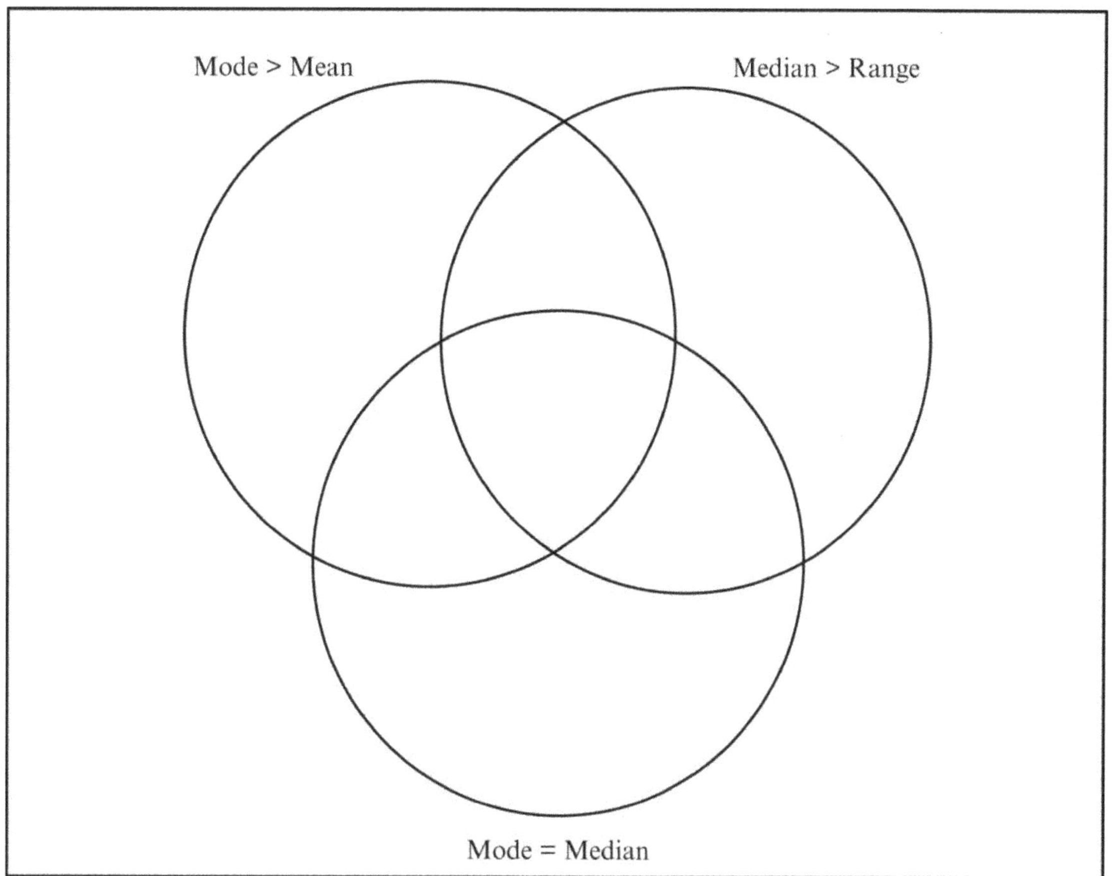